An Illustrated Review of

BASIC CONCEPTS OF CHEMISTRY, THE CELL & TISSUES

An
Illustrated
Review
of
BASIC CONCEPTS
OF CHEMISTRY,
THE CELL, & TISSUES

Glenn F. Bastian

HarperCollins*CollegePublishers*

Executive Editor: Bonnie Roesch
Cover Designer: Kay Petronio

Transferred to digital print on demand, 2002
Printed & Bound by Antony Rowe Ltd, Eastbourne

AN ILLUSTRATED REVIEW OF BASIC CONCEPTS OF CHEMISTRY, THE CELL, AND TISSUES

by Glenn F. Bastian

Library of Congress Cataloging-in-Publication Data

Bastian, Glenn F.
    An illustrated review of basic concepts of chemistry, the cell, &
tissues / Glenn F. Bastian.
      p.   cm.
    Includes bibliographical references.
    ISBN: 0-06-501703-X
    1. Human physiology—Outlines, syllabi, etc. 2. Biochemistry—
Outlines, syllabi, etc. 3. Cytology—Outlines, syllabi, etc.
4. Histology—Outlines, syllabi, etc. I. Title.
QP41.B37   1993
612'.015—dc20

                                            93–1986
                                            CIP

      9 8 7 6

In Memory
of
Frank V. Boyle, M.D.

CONTENTS

List of Topics & Illustrations   ix
Preface   xiii

PART I: BASIC CONCEPTS OF CHEMISTRY, THE CELL, AND TISSUES

Chapter 1: **Structural Organization of the Body**   1
Chapter 2: **Chemistry**   15
Chapter 3: **The Cell**   35
Chapter 4: **Tissues**   49
Chapter 5: **Homeostasis**   65

PART II: SELF-TESTING EXERCISES   73

All illustrations used in Part I are unlabeled for self-testing.

PART III: TERMINOLOGY

Pronunciation Guide   108
Glossary   113
Bibliography   123

LIST OF TOPICS & ILLUSTRATIONS

Text: One page of text is devoted to each of the following topics. *Illustrations are listed in italics.*

Chapter 1: Structural Organization of the Body   1
Levels of Organization   2
  levels of organization
Organ Systems   4
  skeletal system
Regional Names   6
  regional names
Planes and Directional Terms   8
  planes and directional terms
Body Cavities   10
  body cavities and abdominal regions
  mediastinum
  abdominopelvic cavity

Chapter 2: Chemistry   15
Matter   16
  chemical formulas
Chemical Bonds   18
  types of chemical bonds
Chemical Reactions   20
  types of chemical reactions
Water and Electrolytes   22
  water and electrolytes

Organic Compounds   24
organic compounds
Proteins   26
protein structure (insulin)
Acids and Bases   28
the pH scale
Energy   30
ATP
Diffusion, Osmosis, and Bulk Flow   32
diffusion and osmosis

Chapter 3: **The Cell**   35
Cell Structures   36
cell structures
Plasma Membrane   38
membrane transport
Nucleus   40
cell division: mitosis and meiosis
Mitochondria   42
cellular respiration: energy (ATP) production
Ribosomes   44
protein synthesis: mRNA and tRNA
Cell Communication   46
ligands and receptors

Chapter 4: **Tissues**   49
Overview   50
the four basic tissues
Epithelial Tissues   52
covering and lining epithelium: simple and stratified
Connective Tissues   54
the five basic types of connective tissues
Nervous Tissues   56
nervous tissues: neurons and neuroglia
Muscle Tissues   58
muscle tissues: skeletal, cardiac, and smooth muscle

Integumentary System   **60**
   skin
Skin Functions   **62**
   skin receptors

Chapter 5: **Homeostasis**   **65**
Survival: The Unifying Theme   **66**
   survival of individual cells
The Internal Environment   **68**
   fluid compartments
Feedback Systems   **70**
   feedback systems (reflex arcs)

PREFACE

An Illustrated Review of Anatomy and Physiology is a series of ten books written to help students effectively review the structure and function of the human body. Each book in the series is devoted to a different body system.

My objective in writing these books is to make very complex subjects accessible and unthreatening by presenting material in manageable size bits (one topic per page) with clear, simple illustrations to assist the many students who are primarily visual learners. Designed to supplement established texts, they may be used as a student aid to jog the memory, to quickly recall the essentials of each major topic, and to practice naming structures in preparation for exams.

INNOVATIVE FEATURES OF THE BOOK

(1) Each major topic is confined to one page of text.

A unique feature of this book is that each topic is confined to one page and the material is presented in outline form with the key terms in boldface or italic typeface. This makes it easy to quickly scan the major points of any given topic. The student can easily get an overview of the topic and then zero in on a particular point that needs clarification.

(2) Each page of text has an illustration on the facing page.

Because each page of text has its illustration on the facing page, there is no need to flip through the book looking for the illustration that is referred to in the text ("see Figure X on page xx"). The purpose of the illustration is to clarify a central idea discussed in the text. The images are simple and clear, the lines are bold, and the labels are in a large type. Each illustration deals with a well-defined concept, allowing for a more focused study.

> *PHYSIOLOGY TOPICS* (1 text page : 1 illustration)
> Each main topic in physiology is limited to one page of text with one supporting illustration on the facing page.

ANATOMY TOPICS (1 text page : several illustrations)
For complex anatomical structures a good illustration is more valuable than words. So, for topics dealing with anatomy, there are often several illustrations for one text topic.

(3) Unlabeled illustrations have been included.

In Part II, all illustrations have been repeated without their labels. This allows a student to test his or her visual knowledge of the basic concepts.

(4) A Pronunciation Guide has been included.

Phonetic spelling of unfamiliar terms is listed in a separate section, unlike other textbooks where it is usually found in the glossary or spread throughout the text. The student may use this guide for pronunciation drill or as a quick review of basic vocabulary.

(5) A glossary has been included.

Most textbooks have glossaries that include terms for all of the systems of the body. It is convenient to have all of the key terms for one system in a single glossary.

ACKNOWLEGDMENTS

I would like to thank the reviewers of the manuscript for this book who carefully critiqued the text and illustrations for their effectiveness: William Kleinelp, Middlesex County College, Jean Helgeson, Collin County Community College, and Robert Smith, University of Missouri, St. Louis and St. Louis Community College, Forest Park. Their help and advice is greatly appreciated. I am greatly indebted to my editor Bonnie Roesch for her willingness to try a new idea, and for her support throughout this project. I invite students and instructors to send any comments and suggestions for enhancements or changes to this book to me, in care of HarperCollins, so that future editions can continue to meet your needs.

Glenn Bastian

An Illustrated Review of

BASIC CONCEPTS OF CHEMISTRY, THE CELL, & TISSUES

1 Structural Organization of the Body

Levels of Organization *2*
1. Chemical
2. Cellular
3. Tissue
4. Organ
5. System
6. Organismic

Organ Systems *4*
1. Integumentary
2. Skeletal
3. Muscular
4. Nervous
5. Endocrine
6. Cardiovascular
7. Lymphatic
8. Respiratory
9. Digestive
10. Urinary
11. Reproductive

Regional Names *6*
1. Head and Neck
2. Trunk
3. Upper Extremities
4. Lower Extremities

Planes & Directional Terms *8*
1. Planes : sagittal, frontal, transverse, oblique
2. Directional Terms :
   superior — inferior   proximal — distal
   anterior — posterior   superficial — deep
   medial — lateral

Body Cavities *10*
1. Dorsal Body Cavity
   Subdivisions : Cranial Cavity and Vertebral Canal
2. Ventral Body Cavity
   Subdivisions : Thoracic Cavity and Abdominopelvic Cavity
   Viscera
   Serous Membranes

STRUCTURAL ORGANIZATION / Levels of Organization

CHEMICAL LEVEL   *All matter is made up of atoms and molecules.*
   Atom : an atom is an electrically neutral particle; it consists of a nucleus that contains positive protons and neutral neutrons surrounded by negatively charged electrons. There are about 100 different types of atoms. Each type of atom has a different number of protons.
   Element : an element is a substance that contains only one type of atom.
Only 6 elements make up 99% of all living tissue :

(1) Carbon	(3) Nitrogen	(5) Phosphorus
(2) Hydrogen	(4) Oxygen	(6) Sulfur

   Molecule : atoms combine chemically in definite proportions to form molecules. For example, carbohydrate molecules always contain carbon, hydrogen, and oxygen atoms in a 1 : 2 : 1 ratio.
   Compound : a compound is a substance that contains only one type of molecule. Compounds are classified as organic or inorganic. Organic compounds include all complex compounds of carbon; all other compounds are classified as inorganic.
There are 4 principal types of organic compounds found in the body :

(1) Carbohydrates	(3) Proteins
(2) Lipids	(4) Nucleic Acids

CELLULAR LEVEL   *A cell is made up of organic molecules and water molecules.*
   The average cell is 80% water molecules and 20% organic molecules.
   The cell is the basic structural and functional unit of all living things.
   A cell has 4 principal parts :
   (1) Plasma membrane : 50% protein molecules and 50% lipid molecules.
   (2) Cytosol (Intracellular Fluid) : nutrients, soluble proteins, and ions dissolved in water.
   (3) Organelles : nucleus, ribosomes, endoplasmic reticulum, Golgi complex, lysosomes, peroxisomes, mitochondria, cytoskeleton, flagella and cilia, centrioles.
   (4) Inclusions : glycogen (starch), triglycerides (neutral fats), melanin (pigment).

TISSUE LEVEL   *A tissue is made up of cells of the same type.*
   There are 4 principal kinds of tissues :

(1) Epithelial	(3) Muscle
(2) Connective	(4) Nervous

ORGAN LEVEL   *An organ is made up of a variety of tissues.*
   The stomach is an example of an organ. It contains all of the four principal kinds of tissues.

SYSTEM LEVEL   *A system is made up of several organs that perform related functions.*
   There are 11 systems :

(1) Integumentary	(5) Endocrine	(9) Digestive
(2) Skeletal	(6) Cardiovascular	(10) Urinary
(3) Muscular	(7) Lymphatic	(11) Reproductive
(4) Nervous	(8) Respiratory	

ORGANISMIC LEVEL   *An organism contains all of the systems working together.*
   The immune system protects the body from foreign cells and chemicals; the reproductive system perpetuates the species; other systems maintain homeostasis (a stable internal environment).

LEVELS OF ORGANIZATION

Chemical Level
(glucose)

CH₂OH

Cellular Level
(generalized cell)

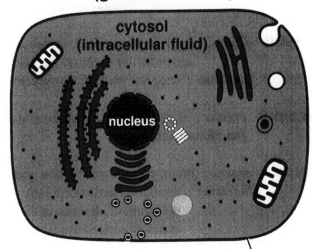

cytosol
(intracellular fluid)

nucleus

plasma membrane

Tissue Level
(epithelium of stomach)

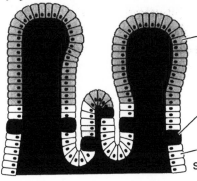

Mucous Cell
secretes mucus

Parietal Cell
secretes HCl

Zymogenic Cell
secretes pepsinogen

System Level
(digestive system)

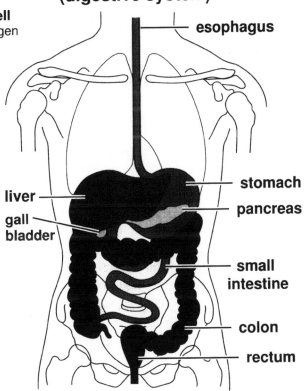

esophagus

stomach

pancreas

liver

gall
bladder

small
intestine

colon

rectum

Organ Level
(stomach wall)

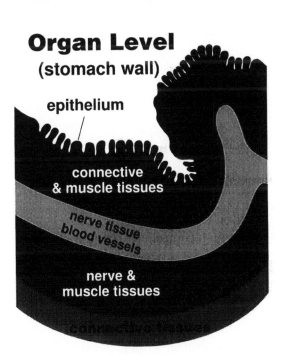

epithelium

connective
& muscle tissues

nerve tissue
blood vessels

nerve &
muscle tissues

STRUCTURAL ORGANIZATION / Organ Systems

Integumentary *organs :* skin and epidermal derivatives (nails, glands, and hair).
　　　　　　functions : sensations (tactile, thermal, pain); protection; immunity; excretion;
　　　　　　　　blood reservoir; vitamin D synthesis; temperature regulation.

Skeletal *organs :* axial skeleton (skull, vertebral column, and ribs);
　　　　　　appendicular skeleton (pectoral & pelvic girdles; upper & lower extremities).
　　　functions : support; protection; movement; blood cell production; mineral storage.

Muscular *organs :* skeletal muscle; cardiac muscle; smooth muscle.
　　　functions : movement.

Nervous *organs :* brain; spinal cord; nerves; ganglia; special sense organs.
　　functions : regulation and coordination of many activities in the body;
　　　　detection of changes in the internal and external environments;
　　　　states of consciousness and learning.

Endocrine *organs :* hypothalamus; anterior pituitary; posterior pituitary; thyroid;
　　　　parathyroid; adrenal cortex; adrenal medulla; kidneys; pancreas;
　　　　stomach; duodenum; testes; ovaries; placenta; thymus; pineal gland.
　functions : regulation and coordination of many activities in the body.

Cardiovascular *organs :* heart; blood vessels; blood.
　　　　functions : transport of food, gases, and hormones to the body's tissues.

Lymphatic *organs :* bone; lymph vessels and nodes; spleen; thymus; lymphoid tissues.
　　　functions : defense against foreign invaders;
　　　　return of extracellular fluid to the blood;
　　　　formation and differentiation of white blood cells.

Respiratory *organs :* nose; pharynx; larynx; bronchi; bronchioles; lungs.
　　　functions : exchange of carbon dioxide and oxygen;
　　　　regulation of hydrogen ion concentration.

Digestive *organs :* mouth; pharynx; esophagus; stomach; intestines; salivary glands;
　　　　pancreas; liver; gallbladder.
　　functions : digestion and absorption of organic nutrients, salts, and water.

Urinary *organs :* kidneys; ureters; urinary bladder; urethra.
　　functions : regulation of plasma composition through controlled excretion of
　　　　organic wastes, salts, and water.

Reproductive *organs :* Male (testes; penis; associated ducts and glands);
　　　　Female (ovaries; uterine tubes; uterus; vagina; mammary glands).
　　　functions : Male (production of sperm; transfer of sperm to female).
　　　　Female (production of eggs; nutritive environment for embryo).

SKELETAL SYSTEM
an example of an organ system

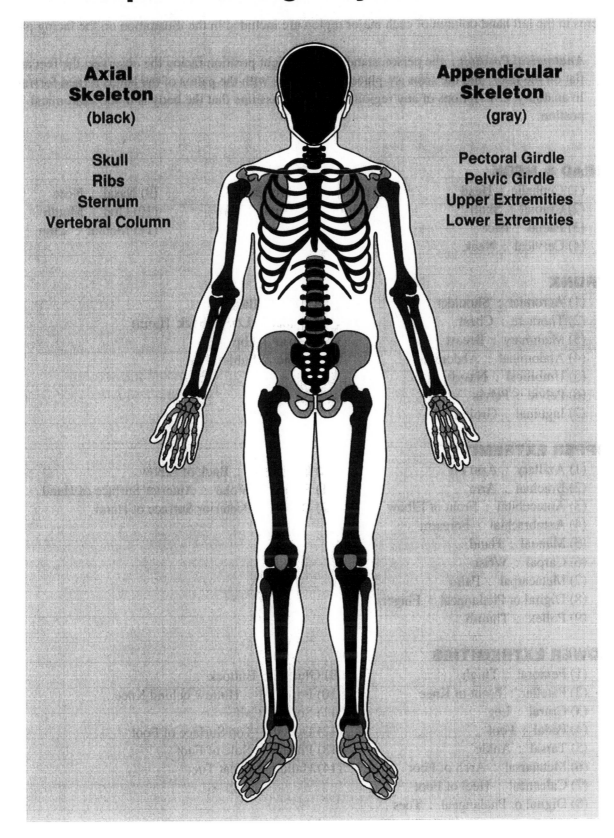

Axial
Skeleton
(black)

Skull
Ribs
Sternum
Vertebral Column

Appendicular
Skeleton
(gray)

Pectoral Girdle
Pelvic Girdle
Upper Extremities
Lower Extremities

STRUCTURAL ORGANIZATION / Regional Names

In the following lists, the anatomical adjective for a region is followed by the common name (noun). Items in the left hand column of each major region are included in the illustration on the facing page.

Anatomical Position : the person stands in an upright position facing the observer; the feet are flat on the floor, and the arms are placed at the sides with the palms of the hands turned forward. In anatomy, descriptions of any region of the body assume that the body is in the anatomical postion.

HEAD & NECK

(1) Cephalic : Head
(2) Cranial : Skull
(3) Facial : Face
(4) Cervical : Neck

(5) Frontal : Forehead
(6) Orbital or Ocular : Eye
(7) Otic : Ear
(8) Buccal : Cheek

(9) Nasal : Nose
(10) Oral : Mouth
(11) Mental : Chin

TRUNK

(1) Acromial : Shoulder
(2) Thoracic : Chest
(3) Mammary : Breast
(4) Abdominal : Abdomen
(5) Umbilical : Navel
(6) Pelvic : Pelvis
(7) Inguinal : Groin

(8) Dorsal : Back
(9) Lumbar : Lower Back (Loin)
(10) Coxal : Hip
(11) Pubic : Pubis

UPPER EXTREMITIES

(1) Axillary : Arm Pit
(2) Brachial : Arm
(3) Antecubital : Front of Elbow
(4) Antebrachial : Forearm
(5) Manual : Hand
(6) Carpal : Wrist
(7) Metacarpal : Palm
(8) Digital or Phalangeal : Fingers
(9) Pollex : Thumb

(10) Olecranial : Back of Elbow
(11) Palmar or Volar : Anterior Surface of Hand
(12) Dorsal : Posterior Surface of Hand

LOWER EXTREMITIES

(1) Femoral : Thigh
(2) Patellar : Front of Knee
(3) Crural : Leg
(4) Pedal : Foot
(5) Tarsal : Ankle
(6) Metatarsal : Arch of Foot
(7) Calcaneal : Heel of Foot
(8) Digital or Phalangeal : Toes

(9) Gluteal : Buttock
(10) Popliteal : Hollow behind Knee
(11) Sural : Calf
(12) Dorsal : Top Surface of Foot
(13) Plantar : Sole of Foot
(14) Hallux : Great Toe

REGIONAL NAMES

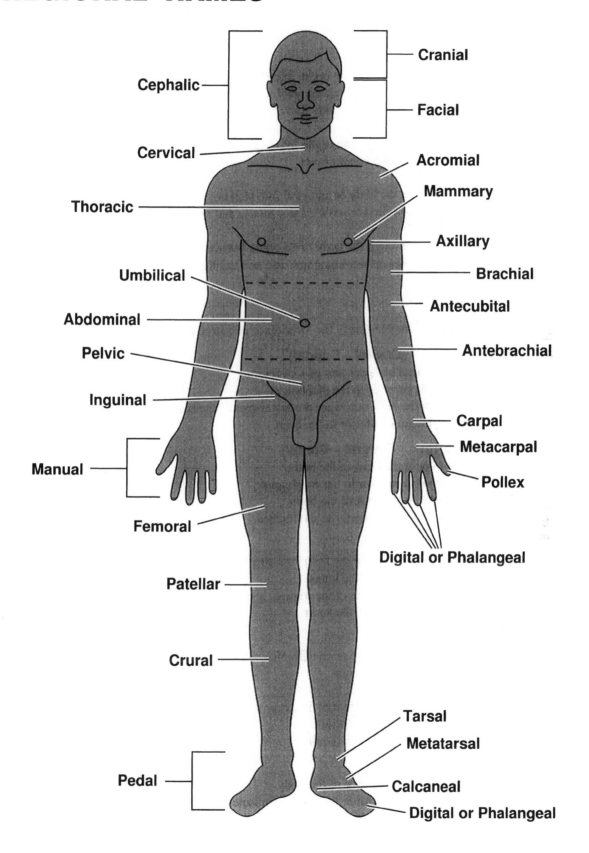

Cephalic

Cranial

Facial

Cervical

Acromial

Mammary

Thoracic

Axillary

Brachial

Umbilical

Antecubital

Abdominal

Antebrachial

Pelvic

Inguinal

Carpal

Metacarpal

Manual

Pollex

Femoral

Digital or Phalangeal

Patellar

Crural

Tarsal

Metatarsal

Pedal

Calcaneal

Digital or Phalangeal

STRUCTURAL ORGANIZATION / Planes & Directional Terms

PLANES

Sagittal Plane
A sagittal plane divides the body or an organ into right and left sides.
> *Midsagittal Plane :* passes through the midline, dividing the body into equal halves.
> *Parasagittal Plane :* divides the body into unequal right and left sides.

Frontal Plane
A frontal plane divides the body or an organ into anterior and posterior portions.
Frontal planes are also called *coronal planes.*

Transverse Plane
A transverse plane divides the body or an organ into superior and inferior portions.
Transverse planes are also called *cross-sectional planes* and *horizontal planes.*

Oblique Plane
An oblique plane passes through the body or an organ between
transverse and frontal planes or between transverse and sagittal planes.

DIRECTIONAL TERMS

Anatomists use directional terms to clarify the location of one structure relative to another.

Superior — Inferior   (Cranial – Caudal)
Superior : toward the head or upper part of a structure.
> The heart is superior to the diaphragm.

Inferior :   away from the head or toward the lower part of the body.
> The diaphragm is inferior to the heart.

Anterior — Posterior   (Ventral – Dorsal)
Anterior :   nearer to or at the front of the body.
> The trachea is anterior to the esophagus.

Posterior : nearer to or at the back of the body.
> The esophagus is posterior to the trachea.

Medial — Lateral
Medial :   nearer to the midline of the body or a structure.
> The heart is medial to the lungs.

Lateral :   farther from the midline of the body or a structure.
> The lungs are lateral to the heart.

Proximal — Distal
Proximal : nearer to the attachment of an extremity to the trunk or a structure;
nearer to the point of origin.
> The shoulder is proximal to the elbow.

Distal :   farther from the attachment of an extremity to the trunk or a structure;
farther from the point of origin.
> The elbow is distal to the shoulder.

Superficial — Deep
Superficial : toward or on the surface of the body.
> The skin is superficial to the skeletal muscles.

Deep : away from the surface of the body.
> The skeletal muscles are deep to the skin.

PLANES AND DIRECTIONAL TERMS

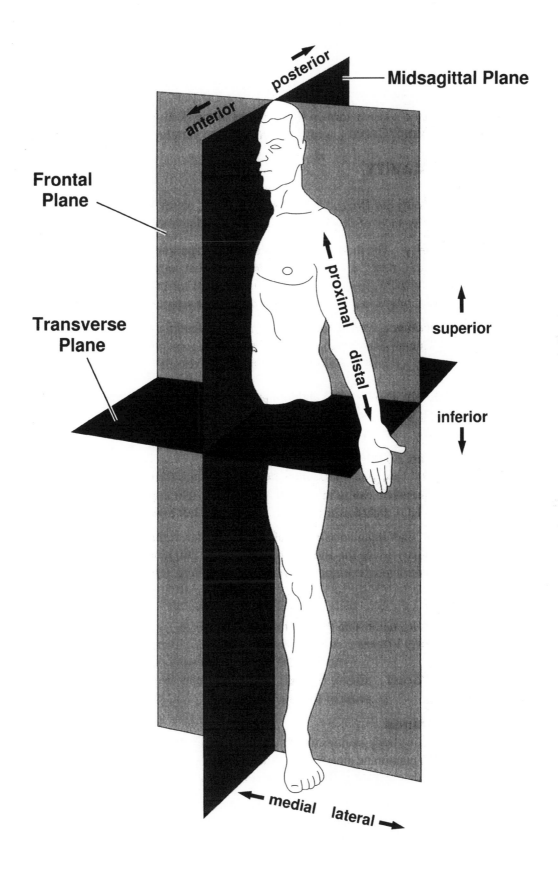

9

STRUCTURAL ORGANIZATION / Body Cavities

Body Cavity : A body cavity is a space within the body that does not open to the exterior. The two principal body cavities are the dorsal body cavity and the ventral body cavity.

DORSAL BODY CAVITY
Subdivisions
Cranial Cavity : a cavity formed by the bones of the skull; it contains the brain.
Vertebral (Spinal) Canal : a cavity formed by the vertebrae; it contains the spinal cord.

VENTRAL BODY CAVITY
Subdivisions
The ventral body cavity has two subdivisions: the thoracic cavity and the abdominopelvic cavity; they are separated by a flat sheet of skeletal muscle called the diaphragm.

Thoracic Cavity   The thoracic cavity has three compartments :
(1) Right Pleural Cavity : a small, fluid-filled space that surrounds the right lung.
(2) Left Pleural Cavity : a small, fluid-filled space that surrounds the left lung.
(3) Pericardial Cavity : a small, fluid-filled space that surrounds the heart.

— *Mediastinum :* a region between the lungs, extending from the sternum (breastbone) to the vertebral column (backbone). Its contents include : esophagus; trachea; bronchi; thymus gland; heart; pericardial cavity; many large blood vessels and lymphatic vessels.

Abdominopelvic Cavity   The abdominopelvic cavity is divided into two portions :
(1) Abdominal Cavity : the portion extending from the diaphragm to the pelvis.
(2) Pelvic Cavity : the portion within the pelvis.

— *Quadrants & Regions*
4 Quadrants : Imaginary horizontal and vertical lines through the umbilicus (navel) divide the abdominopelvic cavity into four quadrants — right upper quadrant (RUQ), left upper quadrant (LUQ), right lower quadrant (RLQ), and left lower quadrant (LLQ).

9 Regions : Four imaginary lines (left midclavicular, right midclavicular, subcostal, and transtubercular) divide the abdominopelvic cavity into nine regions — right hypochondriac, epigastric, left hypochondriac, right lumbar, umbilical, left lumbar, right iliac, pubic, and left iliac.

Viscera   (organs located inside the ventral body cavity)
Abdominal Cavity Viscera : stomach; spleen; pancreas; liver; gall bladder; small intestine; most of the the large intestine.
Pelvic Cavity Viscera : urinary bladder; internal reproductive structures; some of the large intestine.

Serous Membranes
Serous membranes line body cavities and cover the organs within those cavities.
The part of a serous membrane attached to a cavity wall is called the *parietal portion*; the part attached to the organs is called the *visceral portion*.

(1) Pleural Membranes (pleura) : cover the lungs; line the pleural cavities.
(2) Pericardial Membrane (pericardium) : covers the heart; lines the pericardial cavity.
(3) Peritoneal Membrane (peritoneum) : covers the viscera; lines the abdominal cavity.

BODY CAVITIES AND REGIONS

Body Cavities

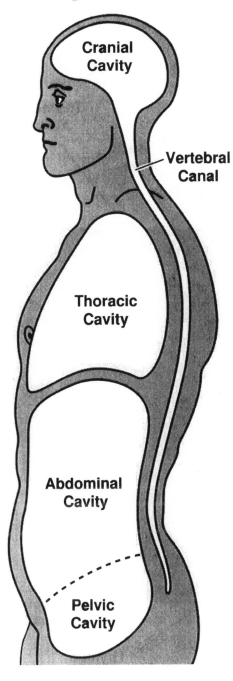

Cranial Cavity

Vertebral Canal

Thoracic Cavity

Abdominal Cavity

Pelvic Cavity

Abdominal Regions

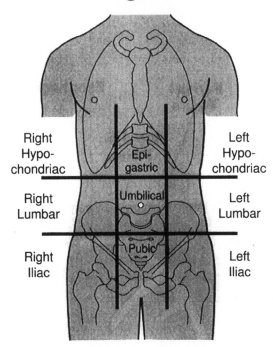

Right Hypo-chondriac

Epi-gastric

Left Hypo-chondriac

Right Lumbar

Umbilical

Left Lumbar

Right Iliac

Pubic

Left Iliac

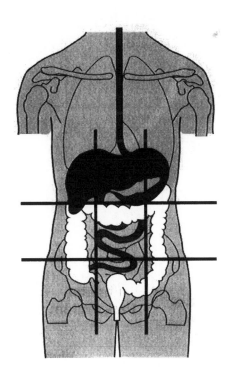

MEDIASTINUM

Transverse section at the level of the 6th thoracic vertebra.

The mediastinum includes all the contents of the thoracic cavity *except* the lungs.

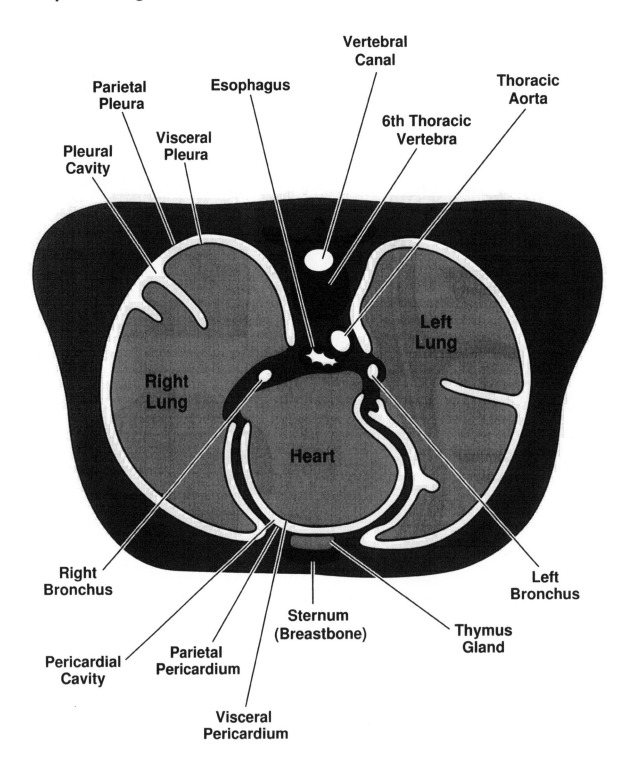

ABDOMINOPELVIC CAVITY
showing peritoneum and viscera

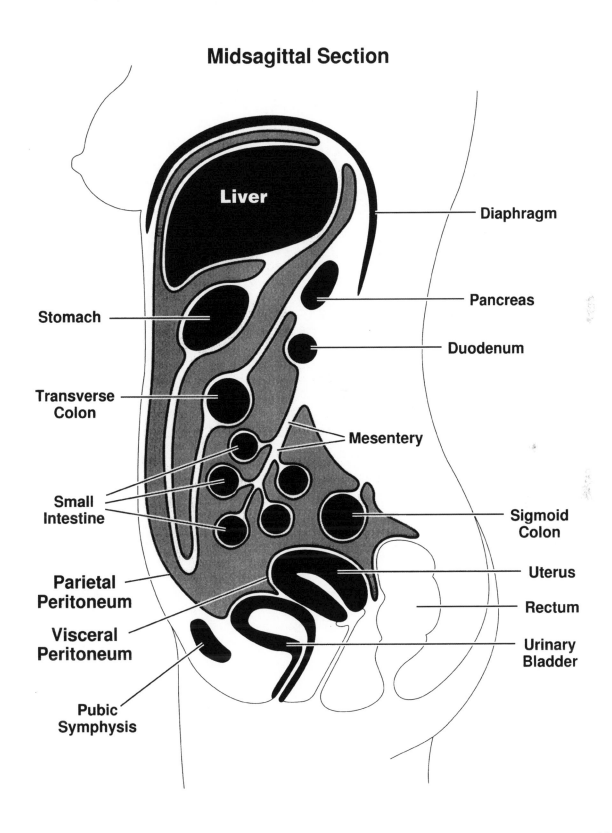

Midsagittal Section

Liver

Diaphragm

Pancreas

Stomach

Duodenum

Transverse
Colon

Mesentery

Small
Intestine

Sigmoid
Colon

Uterus

Parietal
Peritoneum

Rectum

Visceral
Peritoneum

Urinary
Bladder

Pubic
Symphysis

2 Chemistry

Matter *16*
1. Atomic Structure : nucleus, electrons, and related terms
2. Organization of Matter : elements and compounds

Chemical Bonds *18*
1. Electrons
2. Types of Chemical Bonds : covalent, ionic, and hydrogen bonds

Chemical Reactions *20*
1. Chemical Reactions in Cells : anabolism and catabolism
2. Factors that Affect Reaction Rates
3. Related Terms

Water and Electrolytes *22*
1. Water : properties and functions of body water
2. Electrolytes : common electrolytes and ions

Organic Compounds *24*
1. Carbohydrates
2. Lipids
3. Proteins
4. Nucleic Acids

Proteins *26*
1. Amino Acids, Polypeptides, and Proteins
2. Levels of Structural Organization
3. Protein Functions
4. Enzymes

Acids and Bases *28*
1. Acid
2. Base
3. pH

Energy *30*
1. Kinds of Energy
2. Forms of Energy
3. Sources of Chemical Energy (Fuels)
4. ATP (Adenosine Triphosphate)

Diffusion, Osmosis, & Bulk Flow *32*
1. Diffusion
2. Osmosis
3. Bulk Flow
4. Related Terms

CHEMISTRY / Matter

Matter : anything that occupies space and has mass.
Mass : the amount of matter a substance contains.

Weight : the force of gravity acting on a mass.
Density : the mass per unit volume.

ATOMIC STRUCTURE

All matter, living and nonliving, is composed of *atoms*. Atoms are the smallest units of matter that enter into chemical reactions. They are electrically neutral, always having the same number of positively and negatively charged particles. Structurally they have two main parts :

Nucleus

The mass of an atom is concentrated in a tiny nucleus at the center of the atom. The nucleus consists of two types of subatomic particles : positive *protons* and neutral *neutrons*. Protons and neutrons are approximately equal in mass and weigh about 2000 times more than an electron.

Electrons

The space occupied by an atom is determined by the negatively charged electrons that move at enormous speeds in unpredictable pathways around the nucleus. The electrons are organized in *electron shells* that form concentric spheres around the nucleus; the further out the shell is from the nucleus, the higher the energy level of the electrons in the shell.

Related Terms

Atomic Number : the number of protons or electrons in the nucleus.
Atomic Weight (Atomic Mass) : the number of protons plus neutrons.
Ions : atoms that have gained or lost electrons and carry a negative or positive charge.
Isotopes : atoms with the same number of protons but different numbers of neutrons.
        (carbon–12 has 6 protons and 6 neutrons; carbon–14 has 6 protons and 8 neutrons)

ORGANIZATION OF MATTER
Elements

Element : An element is a substance that consists of only one type of atom.
        92 elements occur in nature.
Symbol : each element has a shorthand abbreviation, consisting of the first letter (or first two
        letters) in the English or Latin name of the element (sodium = Na, L. *natrium*).
Essential Elements in the Body : 26 elements are present in the body.
96% of the body's mass : carbon (C), hydrogen (H), oxygen (O), and nitrogen (N).
3.9% of the body's mass : calcium (Ca), phosphorus (P), potassium (K), sulfur (S), sodium (Na), chlorine (Cl),
        magnesium (Mg), iodine (I), and iron (Fe).
0.1% of the body's mass (trace elements) : aluminum (Al), boron (B), chromium (Cr), cobalt (Co), copper (Cu),
fluorine (F), manganese (Mn), molybdenum (Mo), selenium (Se), silicon (Si), tin (Sn), vanadium (V), and zinc (Zn).

Compounds

Compound : A compound consists of two or more different kinds of atoms or ions in definite
        proportions. In the compound water, hydrogen and oxygen atoms are always present in a 2 : 1
        ratio; in sodium chloride, sodium ions and chloride ions are always present in a 1 : 1 ratio.
Molecule : In a molecule, two or more atoms in definite proportions are linked by chemical bonds.
        A molecule of water consists of 2 hydrogen atoms linked to 1 oxygen atom.
Classification of Compounds
        Inorganic Compounds : acids, bases, salts, and water.
        Organic Compounds : carbohydrates, lipids, proteins, and nucleic acids.
Formulas
        Electron-Dot Formulas : indicate the location of the electrons in the outer shells.
        Molecular Formulas : indicate the elements in a molecule and their proportions (H_2O).
        Structural Formulas : indicate the elements in a molecule and the location of covalent bonds.

CHEMICAL FORMULAS
the most abundant elements in a cell

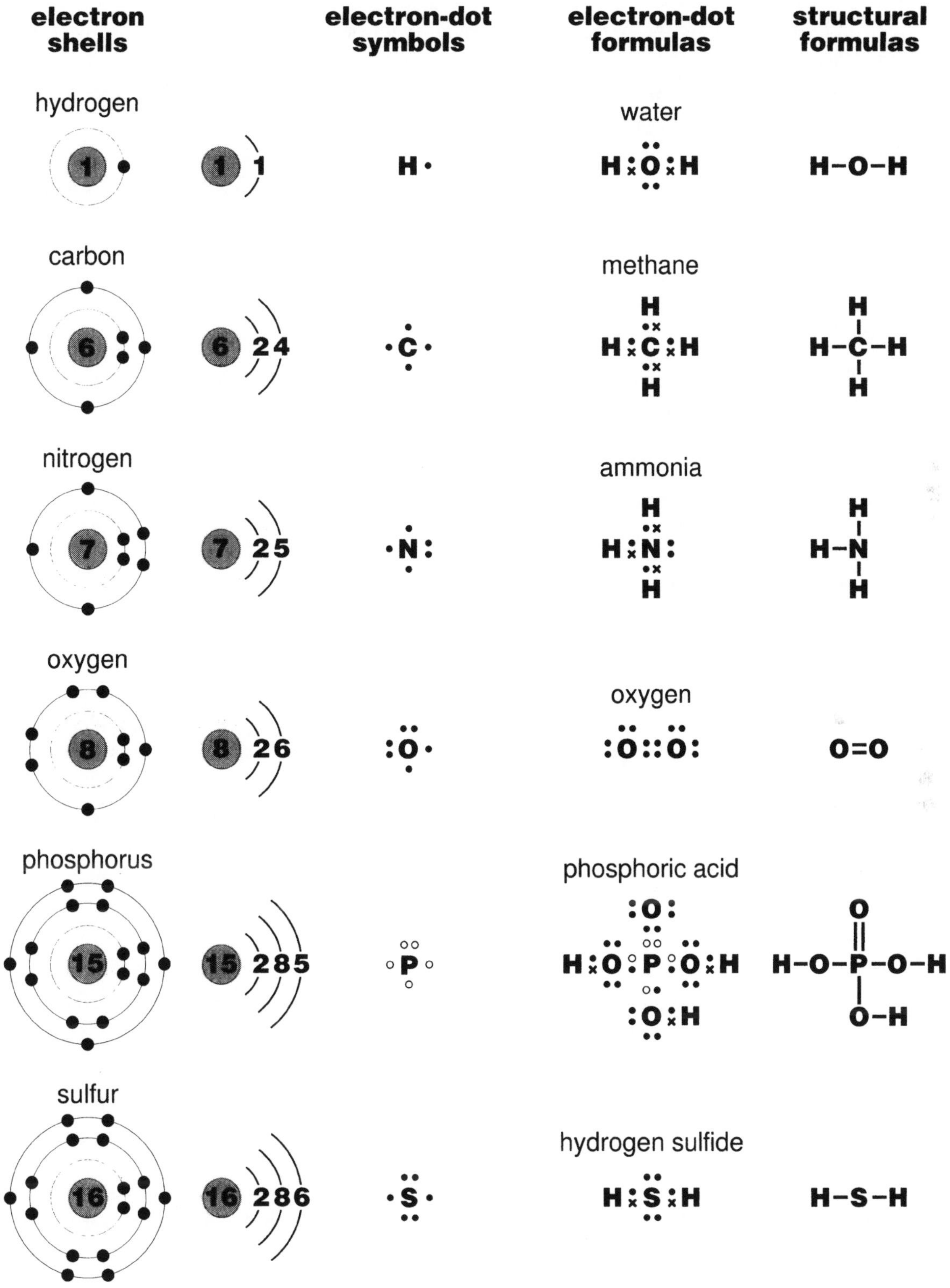

electron shells	electron-dot symbols	electron-dot formulas	structural formulas

hydrogen

H·

water

H:O:H

H–O–H

carbon

·C·

methane

H
H:C:H
H

H
H–C–H
H

nitrogen

·N:

ammonia

H
H:N:
H

H
H–N–H
H

oxygen

:O·

oxygen

:O::O:

O=O

phosphorus

°P°

phosphoric acid

:O:
H:O:P:O:H
:O:H

O
‖
H–O–P–O–H
|
O–H

sulfur

·S·

hydrogen sulfide

H:S:H

H–S–H

CHEMISTRY / Chemical Bonds

ELECTRONS

Electron Interactions
Electron interactions are the basis of all chemical reactions.

Inert Elements
Elements that do not take part in chemical reactions because their outermost electron shells are filled to the maximum are called inert elements. Examples are helium and neon.

Chemical Stability
To achieve chemical stability an atom must fill its outermost electron shell. This is achieved by losing, gaining, or sharing electrons.

Valence Electrons
The electrons in the outermost shell that must be lost, gained, or shared to achieve stability are called the valence electrons.

Electronegativity
Electronegativity is the degree of attraction that an atom has for its electrons. When two atoms form a chemical bond, the difference in their electronegativities determines whether valence electrons are lost, gained, or shared.

> *Highly Electronegative Elements :* oxygen and nitrogen.
> *Moderately Electronegative Elements :* carbon, hydrogen, sulfur, and phosphorus.

TYPES OF CHEMICAL BONDS

(1) Covalent Bonds *(shared electrons)*
When the electronegativities of two combining atoms are close to the same value, valence electrons are shared.

Nonpolar : Bonds formed between two atoms of the same type are called nonpolar covalent bonds, because the electrons forming the bond are shared equally, H—H. Bonds formed between two atoms with very similar electronegativities also share electrons equally, C—H.

Polar : When the electronegativities of two combining atoms are sufficiently different, the electrons forming the bonds are shared unequally—electrons spend more time close to the more electronegative atom, O(–) — (+)H, N(–) — (+)H. Each atom has a partial charge.

(2) Ionic Bonds *(electron transfer)*
If the difference in electronegativity between the combining atoms is great enough, the more electronegative atom will pull the valence electrons away from the less electronegative atom, forming two ions. The ions are held together by electrostatic forces (opposite charges attract).

Electron Acceptor : the atom that accepts one or more electrons and becomes negatively charged.

Electron Donor : the atom that donates one or more electrons and becomes positively charged.

(3) Hydrogen Bonds *(attraction between partially charged atoms)*
Hydrogen bonds occur between molecules that contain polar bonds. The partial *negative* charge associated with a nitrogen or oxygen atom of one molecule interacts with the partial *positive* charge of a hydrogen atom that is part of a second molecule. Hydrogen bonds exist between adjacent water molecules, between polarized regions of the same large molecule (they form within protein molecules), and between the two strands of a DNA molecule.

TYPES OF CHEMICAL BONDS

Covalent Bonds
shared electrons

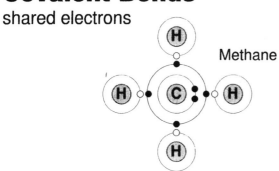

Methane

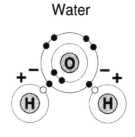

Water

Nonpolar Covalent Bonds
share electrons equally

Polar Covalent Bonds
share electrons unequally

Ionic Bonds
electron transfer

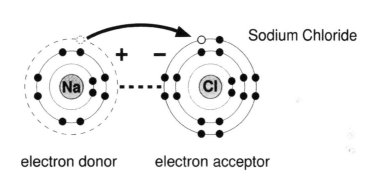

Sodium Chloride

electron donor   electron acceptor

Hydrogen Bonds
partially charged atoms

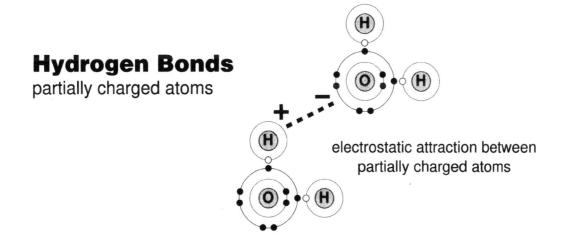

electrostatic attraction between
partially charged atoms

CHEMISTRY / Chemical Reactions

CHEMICAL REACTIONS IN CELLS
Chemical reactions involve breaking chemical bonds and forming new chemical bonds.

Anabolism   (Synthesis Reactions)
Anabolic reactions are synthesis reactions — reactants combine to form a new product.
Anabolic reactions *require* energy.

  Examples:

  Carbohydrate Synthesis : 2 monosaccharides combine to form a disaccharide.
  Lipid Synthesis : 3 fatty acids and 1 glycerol combine to form a triglyceride.
  Protein Synthesis : amino acids combine to form a polypeptide.
  Nucleic Acid Synthesis : nucleotides combine to form DNA or RNA.

Catabolism   (Decomposition Reactions)
Catabolic reactions are decomposition reactions — a molecule breaks into smaller parts.
Catabolic reactions *release* energy.

  Examples :

  Carbohydrate Decomposition : a disaccharide splits, forming 2 monosaccharides.
  Cellular Respiration : a series of 19 reactions that break glucose into CO_2 and H_2O.
  ATP Breakdown : high-energy phosphate bond broken, forming ADP & phosphate.

FACTORS THAT AFFECT REACTION RATES

(1) Temperature
The greater the speed of the particles, the greater the number and force of collisions between the particles. Speed of particles is directly proportional to the temperature (kinetic energy).

(2) Substrate Concentration
The greater the number of particles per unit volume, the greater the number of collisions between the particles.

(3) Enzymes
  Enzyme Concentration : regulated by factors that control the rate of enzyme synthesis.
  Enzyme Activation : enzyme activation is altered by temperature, pH, and cofactors.
  Enzyme Efficiency (Turnover Number) : moles of substrate catalyzed per minute.

RELATED TERMS
Activation Energy : the collision energy (usually in the form of heat) required for a specific reaction to occur. Catalysts (enzymes in metabolic reactions) lower the activation energy.
Catalyst : a substance that alters the rate of a chemical reaction without being changed in the process.
Dehydration Synthesis : formation of a chemical bond with the release of a water molecule.
Enzyme : a substance that alters the rate of a chemical reaction in a cell; an organic catalyst.
Exchange Reaction : a process in which chemical bonds are broken and new bonds are formed; one atom or group of atoms is replaced by another atom or group of atoms.
Hydrolysis : breaking of a chemical bond with the addition of a water molecule.

Metabolic Pathway : a series of reactions by which a substrate is transformed into a product(s).
Metabolism : all of the chemical reactions that occur in cells (anabolic and catabolic reactions).
Metabolite : a molecule that participates in metabolism.
Product : a molecule formed in an enzyme-mediated reaction.
Reversible Reaction : a combination of synthesis and decomposition reactions.
Substrate : a reactant in an enzyme-mediated reaction.

TYPES OF CHEMICAL REACTIONS

Anabolism (Synthesis Reaction)

Dehydration Synthesis (loss of a water molecule)

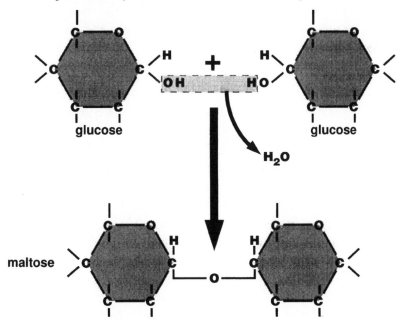

Catabolism (Decomposition Reaction)

Hydrolysis (addition of a water molecule)

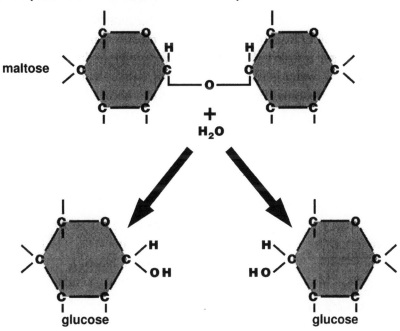

CHEMISTRY / Water and Electrolytes

WATER

Body Water : weight and volume

The most abundant compound in the body is water. It is responsible for about 70% of the total body weight, and it has a volume of 42 liters (1 liter = 1.06 quarts). For a body cell to stay alive it must have a thin layer of water surrounding it. The outer layer of skin cells is not surrounded by fluid; it consists of *dead* cells completely filled with keratin.

Properties of Water

The properties of water that make it so important for the survival of cells are due to its molecular structure : although it is an electrically neutral molecule, it is relatively negative at one end and positive at the other. This is because the oxygen atom has a stronger attraction for electrons than the hydrogen atoms (it is more electronegative). Consequently, the electrons in the covalent bonds are pulled closer to the oxygen atom; thus the bonds are *polar* covalent bonds. The two polar bonds form an angle of 104.5 degrees with the oxygen atom at the corner; because of the orientation of the bonds the water molecule is positive at one end and negative at the other.

The following list summarizes the important properties of water :

(1) Polarity : water molecules have oppositely charged poles.
(2) Hydrogen Bonds : weak attractions between nearby water molecules due to polarity.
(3) Surface Tension, Cohesion, and Adhesion : attractive forces due to hydrogen bonds.
(4) High Specific Heat : 1 calorie of heat is required to raise 1 gram of water 1 degree C.
(5) High Heat of Vaporization : heat required to evaporate water (cooling of sweat).
(6) Solvent : a good solvent for salts and polar molecules.
(7) Ionization of Water : tendency to separate into hydrogen and hydroxide ions;
           pure water has equal numbers of hydrogen and hydroxide ions.

Functions of Body Water

(1) Temperature Moderation : due to high specific heat and high heat of vaporization.
(2) Transport & Exchange : for nutrients, wastes, hormones, antibodies, gases, etc.
(3) Medium for Chemical Reactions : molecules must be dissolved in water to react.
(4) Hydrolysis : water is inserted during the digestive breakdown of foods.
(5) Body Lubricants : water is the base for mucus, saliva, tears, serous and synovial fluids.
(6) Protection : cerebrospinal fluid (brain and spinal cord); amniotic fluid (fetus).

ELECTROLYTES

Electrolytes are substances that dissociate when dissolved in water, releasing charged particles called ions.

Common Electrolytes that dissociate in Body Fluids

Sodium Chloride : $NaCl$	Calcium Carbonate : $CaCO_3$
Potassium Chloride : KCl	Calcium Phosphate : $Ca_3(PO_4)_2$
Calcium Chloride : $CaCl_2$	Sodium Sulfate : Na_2SO_4
Magnesium Chloride : $MgCl_2$	Sodium Bicarbonate : $NaHCO_3$

Common Ions found in Body Fluids

Cations (positive ions) : sodium (Na^+); potassium (K^+); calcium (Ca^{+2}); magnesium (Mg^{+2})
Anions (negative ions) : chloride (Cl^-); carbonate (CO_3^-); phosphate (PO_4^{-3}); sulfate (SO_4^{-2})

WATER AND ELECTROLYTES

Polarity of Water Molecules

Polar Covalent Bonds

$$O$$
$$H^+ \quad {}^{-} \quad {}^{-} \quad {}^+ H$$

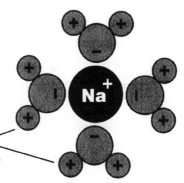

Hydrated Sodium Ion

Na^+

water molecules

Electrolytes ⟶ **Cations** | **Anions**

dissolved in water

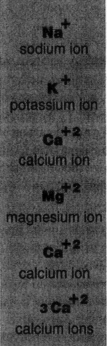

SALTS

NaCl ⟶ Na^+ Cl^-
sodium chloride sodium ion chloride ion

KCl ⟶ K^+ Cl^-
potassium chloride potassium ion chloride ion

CaCl$_2$ ⟶ Ca^{+2} $2Cl^-$
calcium chloride calcium ion chloride ions

MgCl$_2$ ⟶ Mg^{+2} $2Cl^-$
magnesium chloride magnesium ion chloride ions

CaCO$_3$ ⟶ Ca^{+2} CO_3^{-2}
calcium carbonate calcium ion carbonate ions

Ca$_3$(PO$_4$)$_2$ ⟶ $3Ca^{+2}$ $2PO_4^{-3}$
calcium phosphate calcium ions phosphate ions

Na$_2$SO$_4$ ⟶ $2Na^+$ SO_4^{-2}
sodium sulfate sodium ions sulfate ion

ACIDS

HCl ⟶ H^+ Cl^-
hydrochloric acid hydrogen ion chloride ion

BASES

NaHCO$_3$ ⟶ Na^+ HCO_3^-
sodium bicarbonate sodium ion bicarbonate ion

CHEMISTRY / Organic Compounds

Elements : C, H, N, O, P, S. All organic compounds contain carbon (C) and hydrogen (H). Many also contain nitrogen (N) and oxygen (O); some contain phosphorus (P) and sulfur (S).

Carbon Framework : Carbon atoms react with other carbon atoms forming chains and rings. Thousands of different molecules can be formed from these carbon skeletons.

Polymers and Monomers: Large molecules (macromolecules) formed by linking many subunits together are called polymers; the subunits of polymers are called monomers.

CARBOHYDRATES
Carbohydrates are sugars or starches; they provide most of the energy used by cells.
   Monosaccharides : simple sugars including glucose, fructose, and galactose.
   Disaccharides : maltose (2 glucose); sucrose (glucose & fructose); lactose (glucose & galactose).
   Polysaccharides : plant starch, animal starch (glycogen), and cellulose.

LIPIDS
Lipids are a diverse group of compounds that are nonpolar, and therefore insoluble in water.
   Triglycerides (Neutral Fats) : the most common lipids in the body. A triglyceride molecule consists of 1 glycerol and 3 fatty acids. All excess food is converted into triglycerides and stored in fat cells (adipose tissue) as an energy reserve. Adipose tissue insulates and protects the organs.
      Saturated Fats : saturated with hydrogen atoms; contain no double bonds.
      Unsaturated Fats : not saturated with hydrogen atoms; contain double bonds.
   Phospholipids : consist of 1 glycerol, 2 fatty acids, and 1 phosphate; membrane component.
   Steroids : consist of 4 rings of carbon atoms; include cholesterol, some hormones, bile salts.
   Eicosanoids : derived from arachidonic acid; includes prostaglandins and leukotrienes.
   Fat-Soluble Vitamins : vitamins A, D, E, and K.
   Carotenes : yellow pigments found in egg yolk, carrots, and tomatoes; needed for vitamin A.

PROTEINS
   Amino Acids Amino acids are the building blocks (or monomers) of proteins. There are 20 different amino acids which can be linked together in different sequences to form thousands of different protein molecules. Each amino acid has 4 main parts : a central carbon atom, an amino group ($-NH_2$), a carboxyl group ($-COOH$), and an R group (a variable side chain). Each of the 20 amino acids has a different R group. A *peptide bond* is a covalent bond linking the carboxyl group of one amino acid to the amino group of another amino acid.
   Polypeptides A polypeptide is a polymer consisting of 10 to over 2000 amino acids.
   Proteins A protein consists of one to several polypeptides.

NUCLEIC ACIDS
   Nucleotides Nucleotides are the building blocks (monomers) of nucleic acids. There are 4 different types of nucleotides in a given nucleic acid. Each nucleotide has 3 main parts : a nitrogenous base, a phosphate group, and a sugar.
   DNA (deoxyribonucleic acid) DNA is the nucleic acid found in the nucleus of cells. The hereditary information for the entire organism is stored in the DNA of every cell. A gene is a sequence of DNA nucleotides that serves as a blueprint for the synthesis of a particular polypeptide (portion of a protein).
   RNA (ribonucleic acid) There are 3 types of RNA. *Messenger RNA (mRNA)* carries information needed for polypeptide synthesis from the nucleus to ribosomes (cell structures that provide the sites for protein synthesis); mRNAs also serve as templates (patterns) for building specific amino acid sequences. *Transfer RNA (tRNA)* carries specific amino acids to the messenger RNA attached to ribosomes. Ribosomes are made of *ribosomal RNA (rRNA)* and protein.

ORGANIC COMPOUNDS

Carbohydrates and lipids consist of carbon, hydrogen, and oxygen atoms. Polysaccharides are polymers made up of chains of simple sugars such as glucose.

The main types of lipids are triglycerides, phospholipids, steroids, lipoproteins, and eicosanoids.

Carbohydrates
glucose $C_6H_{12}O_6$

Lipids
a triglyceride

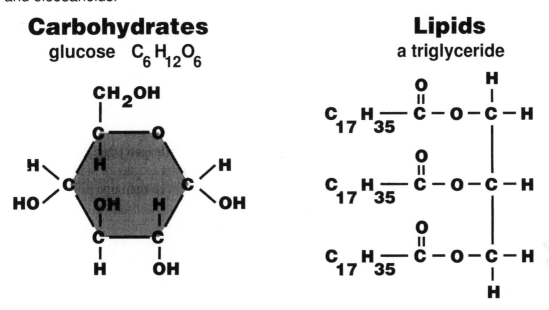

Proteins and nucleic acids contain nitrogen atoms (as well as carbon, hydrogen, & oxygen atoms), which distinguishes them from carbohydrates and lipids.

Proteins consist of chains of amino acids.
Nucleic acids consist of chains of nucleotides.

Nucleic Acids
a nucleotide

Proteins
an amino acid

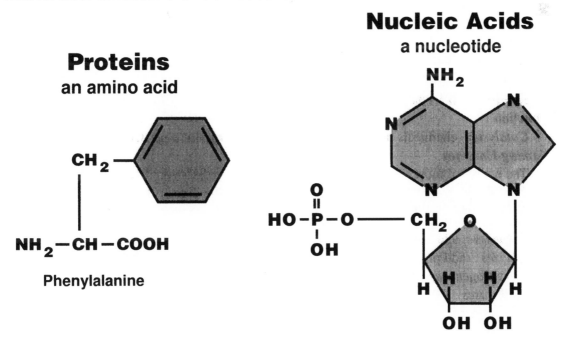

Phenylalanine

CHEMISTRY / Proteins

AMINO ACIDS, POLYPEPTIDES, and PROTEINS
Amino Acids : the building blocks of proteins; there are 20 different types.
Polypeptide : 10 to 2000 amino acids linked by covalent bonds.
Protein : 1 polypeptide or several polypeptides linked by covalent bonds.
    Examples : *Lysozyme (enzyme) :* 1 polypeptide, consisting of 129 amino acids.
                 Insulin (hormone) : 2 polypeptides; a total of 51 amino acids.
                 Collagen (fiber) : 3 polypeptides; a total of about 1,000 amino acids.
                 Hemoglobin : 4 polypeptides; a total of about 600 amino acids.

LEVELS OF STRUCTURAL ORGANIZATION
(1) Primary : the sequence of amino acids in a polypeptide.
(2) Secondary : the spiral or pleated shape of a polypeptide (hydrogen bonding; regular intervals).
(3) Tertiary : the three-dimensional shape of a polypeptide (interactions of side chains).
(4) Quaternary : the shape resulting from the interactions of more than one polypeptide.

PROTEIN FUNCTIONS
(1) Structural : collagen fibers are found in connective tissues; keratin is in hair and skin.
(2) Regulatory : many hormones with a wide variety of regulatory functions are proteins.
(3) Contractile : actin and myosin are protein filaments found in all muscle cells.
(4) Immunological : antibodies and interleukins are proteins.
(5) Transport : hemoglobin carries oxygen and carbon dioxide; lipoproteins carry lipids.
(6) Catalytic : enzymes; catalysts alter the rate of a reaction without changing themselves.

ENZYMES
Parts
    Protein Portion (Apoenzyme) : the main part of the enzyme molecule.
    Nonprotein Portion (Cofactor) : metal ions or coenzymes (vitamin derivatives).
    Active Site : region of an enzyme that has a shape that fits a portion of the substrate molecule.
Shape
    Specificity : each type of enzyme has a unique three-dimensional shape.
    Lock & Key Theory : a substrate fits into the active site of an enzyme as a key fits a lock.
    Denaturation : changes in pH or temperature can change enzyme shape, affecting function.
Function
    Catalysts : chemicals that lower the activation energy for a particular reaction.
Naming Enzymes
    The names of most enzymes end in —*ase.* Enzymes discovered before this system of naming
    was established do not end in —*ase;* an example is the digestive enzyme *pepsin.*
    Enzymes are grouped according to the types of reactions they catalyze :

Oxidases : add oxygen   *Transferases :* transfer groups of atoms
Kinases : add phosphate   *Hydrolases :* add water
Dehydrogenases : remove hydrogen   *Proteases :* break down proteins
Isomerases : rearrange atoms   *Lipases :* break down lipids

PROTEIN STRUCTURE

Insulin
(51 amino acids)

A Chain

B Chain

Disulfide Bond

Amino Acids (abbreviations):

Ala = alanine
Arg = arginine
Asn = asparagine
Asp = aspartic acid
Cys = cystine
Gln = glutamine
Glu = glutamic acid
Gly = glycine

His = histidine
ILE = isoleucine
Leu = leucine
Lys = lysine
Phe = phenylalanine
Pro = proline

Ser = serine
Thr = threonine
Tyr = tyrosine
Val = valine

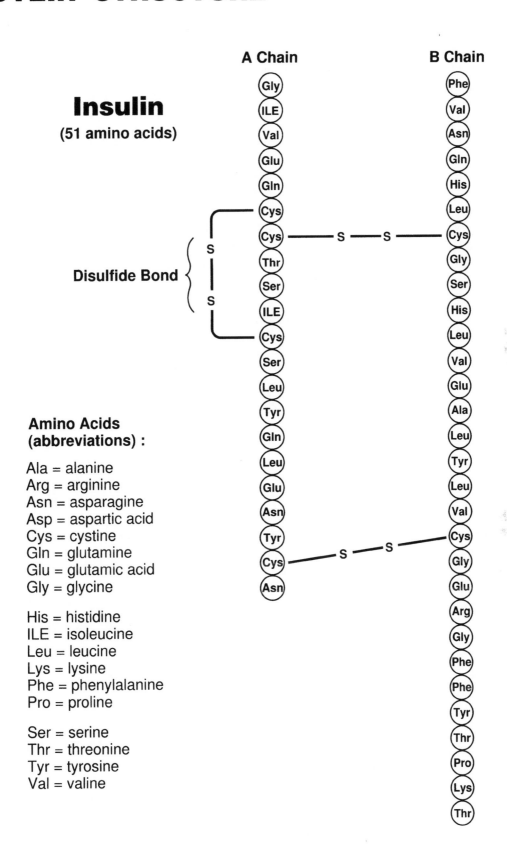

CHEMISTRY / Acids and Bases

ACID

Defined : a compound which, when it dissociates in water, liberates hydrogen ions.

Dissociation is the breaking apart of a molecule to form ions. For example, when hydrogen chloride is dissolved in water, most of the hydrogen chloride molecules break apart, forming hydrogen ions and chloride ions.

Strong and Weak Acids

In strong acids such as hydrochloric acid (HCl) most of the molecules dissociate, producing many hydrogen ions; in weak acids such as carbonic acid (H_2CO_3) many of the molecules do not dissociate, so fewer hydrogen ions are released.

Inorganic Acids

Carbonic Acid : results when metabolically produced carbon dioxide combines with water.

Hydrochloric Acid : secreted by specialized cells lining the stomach wall.

Sulfuric Acid : results from the oxidation of sulfur-containing amino acids.

Phosphoric Acid : results from the oxidation of phosphorus-containing amino acids or the oxidation of nucleic acids (all nucleic acids contain phosphorus).

Organic Acids

Fatty Acids : a major component of triglycerides (neutral fats) and phospholipids.

Ketone Bodies : result from the incomplete oxidation of fatty acids.

Proteins : the carboxyl end (— COOH) can release a hydrogen ion, forming — COO^- + H^+.

BASE

A base is any molecule which can combine with hydrogen ions. Examples of inorganic bases found in the body are sodium bicarbonate and sodium monohydrogen phosphate. Both are important components of buffer systems that keep the pH of body fluids relatively stable. All proteins can act as bases; the amino end (— NH_2) of a polypeptide can combine with a hydrogen ion, forming — NH_3^+.

pH

Concentration of Particles in Solution (Moles)

The concentration of microscopic particles in solution is measured in units called moles.

1 mole = 6.02×10^{23} particles (atoms, molecules, or ions).

pH Defined : pH measures the concentration of hydrogen ions in moles per liter.

p stands for the power of 10 and *H* stands for hydrogen ions.

pH 2 = 1.0×10^{-2} moles per liter of hydrogen ions (or 1/100 moles per liter).

A solution with pH 2 has 10 times as many hydrogen ions as a solution with pH 3.

pH Scale (pH 0 — pH 14)

above 7 = base (pH 7.35 — 7.45 is normal for blood plasma).

pH 7 = neutral (pure water).

below 7 = acid (pH 2 is normal for gastric juices in the stomach).

Buffer Systems

Chemicals that keep the pH of the body fluids relatively constant are called buffers. Buffer systems convert *strong* acids or bases into *weak* acids or bases. Weak acids or bases have little effect on the pH. The principal buffer systems are the bicarbonate buffer system, the phosphate buffer system, and the protein buffer system.

Effect of pH on Enzymes

Most reactions in the body are regulated by enzymes. A change in the pH can alter the shape of an enzyme, which alters its effectiveness as a catalyst. Digestive enzymes in the stomach function best when the pH is 2, while those in the small intestine require a pH of about 8.

pH SCALE

As the acidity of a solution increases, the pH decreases

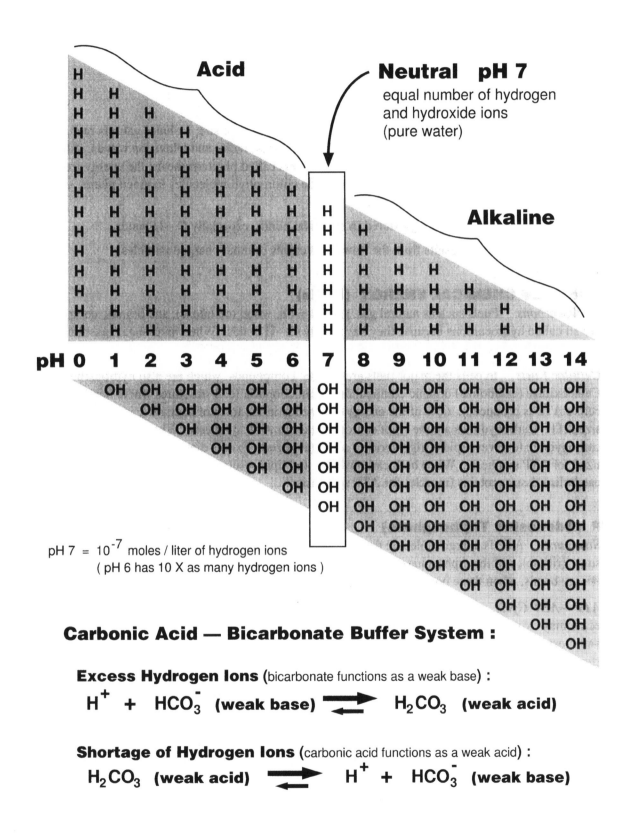

Acid

Neutral pH 7
equal number of hydrogen
and hydroxide ions
(pure water)

Alkaline

| pH | 0 | 1 | 2 | 3 | 4 | 5 | 6 | 7 | 8 | 9 | 10 | 11 | 12 | 13 | 14 |

pH 7 = 10^{-7} moles / liter of hydrogen ions
(pH 6 has 10 X as many hydrogen ions)

Carbonic Acid — Bicarbonate Buffer System :

Excess Hydrogen Ions (bicarbonate functions as a weak base) :

$$H^+ + HCO_3^- \text{ (weak base)} \rightleftarrows H_2CO_3 \text{ (weak acid)}$$

Shortage of Hydrogen Ions (carbonic acid functions as a weak acid) :

$$H_2CO_3 \text{ (weak acid)} \rightleftarrows H^+ + HCO_3^- \text{ (weak base)}$$

29

CHEMISTRY / Energy

Energy : the capacity to do work.

KINDS OF ENERGY
(1) Potential Energy : inactive or stored energy; energy of position.
(2) Kinetic Energy : energy of motion.

FORMS OF ENERGY
(1) Radiant Energy (Electromagnetic Energy) : energy from the sun; includes gamma rays, x-rays, ultraviolet rays, visible light rays, infrared rays, microwaves, radio, and television waves. All chemical energy is derived from radiant energy by the process called photosynthesis; the energy from certain wavelengths of visible light is trapped by the chlorophyll molecules in green plants, and is used to make the C—H bonds of glucose.

(2) Chemical Energy : the energy stored in chemical bonds, especially C—H bonds.

(3) Electrical Energy : results from the flow of electrons or other charged particles.

SOURCES OF CHEMICAL ENERGY (Fuels)
Hydrocarbons   Fuels such as natural gas, gasoline, oil, wood (cellulose), and organic compounds are all called hydrocarbons because they have many C—H bonds. When natural gas, gasoline, oil or wood is burned, C—H bonds are broken and energy is released as heat.

Cellular Fuels   In cells the major fuels are organic compounds, which are also hydrocarbons. Cells oxidize (catabolize) organic compounds in mitochondria (cell structures where nearly all of the cells's ATP is produced). 60% of the energy released is in the form of heat; 40% is used to synthesize ATP. Carbohydrates are the chief source of fuel for ATP synthesis. When there are insufficient carbohydrate reserves, lipids (triglycerides) are catabolized. Excess dietary proteins are also catabolized for ATP synthesis. When both carbohydrate and lipid reserves are exhausted the body will catabolize tissue proteins (muscle) for ATP synthesis.

ATP (Adenosine Triphosphate)
Structure   ATP is a large molecule that consists of a nitrogenous base (adenine), a five-carbon sugar (ribose), and three phosphate groups. The three phosphate groups are linked by two high-energy bonds; when these bonds are broken (hydrolysis of ATP), energy is released.

ATP—ADP Cycle   Energy is continuously cycled through ATP molecules. A typical ATP molecule may exist for only a few seconds before it is broken down into ADP (adenosine diphosphate) and inorganic phosphate; the released energy is used to perform a cell function. The total energy stored in all the ATP molecules of a cell can supply the energy requirements of that cell for less than a minute. Using energy derived from the catabolism of organic compounds, ADP and inorganic phosphate are quickly converted back into ATP. ATP is synthesized by a series of reactions called cellular respiration.

Uses   ATP is the cell's chief energy "currency." It is a form of energy that is immediately available — like cash or a credit card. It is used by cells to carry out their energy-requiring functions, such as the movement of cilia, the contraction of muscle cells, the active transport of molecules across a cell membrane, and the synthesis of organic molecules.

ATP (adenosine triphosphate)

Energy-dependent activities of cells depend upon the energy stored in the high-energy phosphate bonds of ATP.

High-energy phosphate bonds are broken as energy is required by the cell.

Catabolism of fuels provides energy for the regeneration of ATP from ADP and inorganic phosphates.

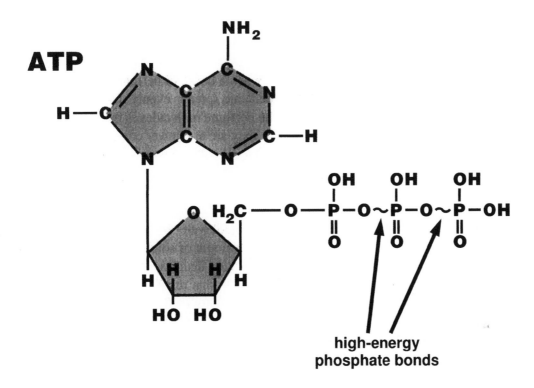

high-energy
phosphate bonds

ATP — ADP Cycle

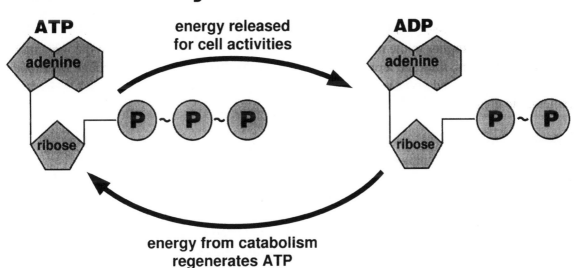

CHEMISTRY / Diffusion, Osmosis, and Bulk Flow

Brownian Motion   Brownian motion is the random motion of microscopic particles suspended in a liquid or gas. It is caused by collisions with molecules of the surrounding medium.   At body temperature water molecules move at about 1500 miles per hour. The higher the temperature, the faster the molecules move. Microscopic particles, such as yeast cells, are jostled by the random bombardments of these fast moving water molecules.

DIFFUSION

Concentration Gradient : a difference in concentration of particles (molecules or ions) in two regions of a fluid (gas or liquid).

Due to the random movement of molecules and the collisions that result, molecules of a given type will eventually be uniformly distributed throughout a system.   The process of diffusion can be observed when ink is dropped into a glass of water or when a bottle of perfume is opened in a room; the ink spreads evenly throughout the water and the perfume spreads evenly throughout the air in the room. The net movement of the ink molecules and the perfume molecules is from a region where they are comparatively concentrated to a region where they are at a lower concentration; in other words, they move <u>down</u> their concentration gradients. Equilibrium is reached when the molecules are evenly distributed throughout the system.

OSMOSIS

Osmosis is a special kind of diffusion: it is the diffusion of water molecules through a semiper- meable membrane. Water concentration is altered by the amount of solute particles. For example, if salt or sugar is dissolved in pure water, the water concentration decreases since some of the space is now occupied by the dissolved molecules of sugar or the ions that result from the dissociation of the salt. If a glucose solution is separated from pure water by a membrane that is permeable to water molecules but not permeable to the larger glucose molecules, water will diffuse from the region of higher concentration (pure water), through the membrane, to the region of lower concentration (glucose solution); this is an example of osmosis.

BULK FLOW

Pressure Gradient : a difference in the hydrostatic pressures at two ends of a fluid-filled tube or between the inside and the outside of the tube.

Bulk flow is the movement of a fluid (liquid or gas) from a region of higher pressure to one of lower pressure. It applies to the movement of a mass of molecules rather than to the movement of individual molecules; thus, the term "bulk." In the human body blood flows through the blood vessels by bulk flow and air flows in and out of the lungs by bulk flow. The greater the pressure gradient, the greater the rate of flow.

Related Terms

Solution : a liquid containing dissolved molecules or ions.
Solute : the molecules or ions dissolved in a liquid.
Solvent : the liquid (water in cells) in which molecules or ions are dissolved .
Hypotonic Solution : one which has a lower solute concentration than the cytosol (cell fluid).
Hypertonic Solution : one which has a higher solute concentration than the cytosol.
Isotonic Solution : one whose water concentration is the same as that of the cytosol.

DIFFUSION AND OSMOSIS

Diffusion

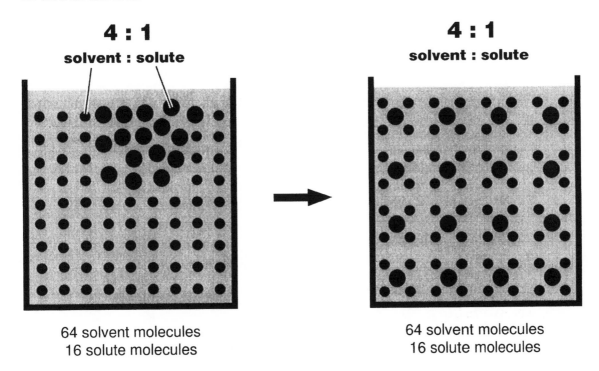

4 : 1
solvent : solute

64 solvent molecules
16 solute molecules

4 : 1
solvent : solute

64 solvent molecules
16 solute molecules

Osmosis : the diffusion of a solvent (usually water) through a semipermeable membrane

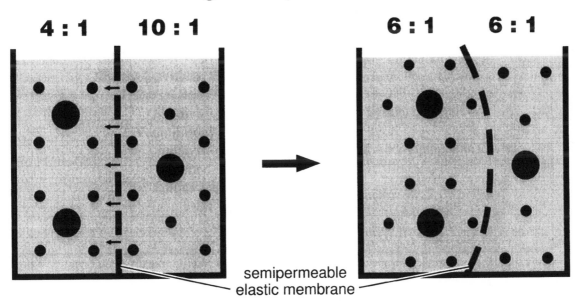

4 : 1   10 : 1

6 : 1   6 : 1

semipermeable
elastic membrane

3 The Cell

Cell Structures *36*

1. Plasma Membrane
2. Nucleus
3. Mitochondria
4. Free Ribosomes
5. Endoplasmic Reticulum
6. Golgi Complex
7. Lysosomes
8. Peroxisomes
9. Cytoskeleton
10. Cilia and Flagella
11. Centrosome and Centrioles
12. Inclusions

Plasma Membrane *38*

1. Structure : lipids and proteins
2. Membrane Transport
   Passive Processes
   Active Processes

Nucleus *40*

1. Structure
2. Cell Division
   Somatic Cell Division : mitosis and cytokinesis
   Reproductive Cell Division : meiosis I and meiosis II

Mitochondria *42*

1. Structure
2. Energy Production
3. Cellular Respiration
   Anaerobic Respiration
   Aerobic Respiration

Ribosomes *44*

1. Structure
2. Protein Synthesis
   Transcription : mRNA synthesis
   Translation : polypeptide synthesis

Cell Communication *46*

1. Ligands and Receptors
2. Receptor Modulation
3. Hormone Receptors

THE CELL / Cell Structures

The cell is the basic structural and functional unit of all living things. Every cell is surrounded by a plasma membrane and is divided into two regions : the *nucleus* and the *cytoplasm*. The cytoplasm is composed of *organelles* (specialized structures) suspended in a fluid called the *cytosol*.

Plasma Membrane   The plasma membrane forms the outer surface of a cell, separating the contents of the cell from the extracellular fluid; it is composed primarily of phospholipids and proteins. *Membrane pores* (also called ion channels) regulate the movement of ions in and out of the cell. *Membrane receptors* (proteins on the outer surface of the membrane) allow the cell to "recognize" other cells of the same type and interact with hormones, enzymes, and antibodies.

Nucleus   The nucleus is a large spherical or oval organelle surrounded by a double membrane; it is usually the most visible organelle. It contains the hereditary information (genes) and controls the activities of the cell by regulating protein synthesis.

Mitochondria (singular : mitochondrion)   Mitochondria are rod-shaped or oval-shaped organelles consisting of a smooth outer membrane and a folded inner membrane. Most of the ATP (energy) used by cells is produced in the mitochondria.

Free Ribosomes   Ribosomes are tiny spheres about 200 nanometers in diameter. They provide the site for protein synthesis; genetic information carried from the nucleus to the ribosomes via messenger RNA determines the sequence of amino acids for a specific protein. Proteins synthesized by free ribosomes perform their functions within the cell (not secreted).

Endoplasmic Reticulum   Reticulum means *network*. Endoplasmic reticulum is an extensive system of flattened sacs or branching tubes. *Smooth (agranular) endoplasmic reticulum* synthesizes steroids and fatty acids. *Rough (granular) endoplasmic reticulum* has ribosomes embedded in the outer surface, giving it a rough appearance; these ribosomes synthesize proteins to be secreted.

Golgi Complex (Apparatus)   The Golgi complex is a stack of 4 to 8 membranous sacs located near the nucleus; proteins and lipids are packaged for distribution in the cell or export from the cell.

Lysosomes   Lysosomes are spherical structures surrounded by a single membrane; they contain digestive enzymes that break down bacteria and debris from dead cells. Large numbers found in white blood cells.

Peroxisomes   Peroxisomes are similar to lysosomes, but smaller. They contain enzymes for the production of *hydrogen peroxide*; *catalase*, another enzyme found in peroxisomes, uses the hydrogen peroxide to detoxify potentially harmful substances.

Cytoskeleton   The cytoskeleton is a network of filaments associated with movement and cellular shape. There are 3 main types of filaments : microfilaments, intermediate filaments, and microtubules.

Cilia and Flagella   Cilia are numerous, short, hairlike structures that extend out from the surface of some epithelial cells. They have a central core of microtubules that produce movement. Their function is to move substances along the surface of the cell. Flagella are longer than cilia and occur singly or in pairs; the only example in the human body is the tail of a sperm cell.

Centrosome and Centrioles   The centrosome is a dense area of cytoplasmic material near the nucleus that contains a pair of cylindrical structures called centrioles; during cell division centrioles generate spindle fibers, which are responsible for the separation of chromosomes.

Inclusions   Inclusions are substances produced by cells and stored in the cytosol. Examples include : *triglycerides*, which are stored in adipose tissue, adrenal cortex cells, and liver cells; *glycogen*, which is stored in muscle and liver cells; and *melanin*, a pigment stored in cells of the skin, hair, and eyes.

CELL STRUCTURES
Generalized Animal Cell

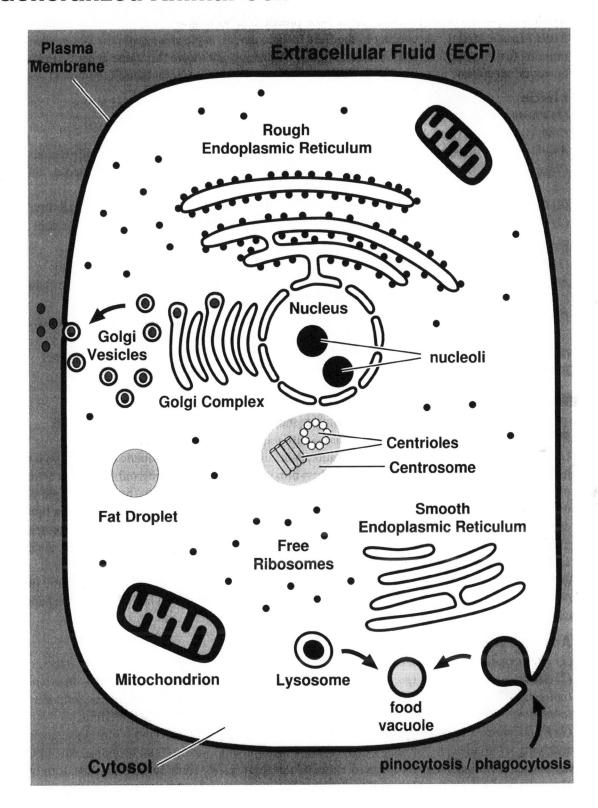

THE CELL / Plasma Membrane

STRUCTURE
By weight, plasma membranes are about 50% protein and 50% lipid.

Fluid Mosaic Model   According to the fluid mosaic model, a plasma membrane is a mosaic of proteins floating like icebergs in a sea of lipids. All membranes have this same general structure; however, membrane proteins vary in different cell types and in different membrane-bound organelles.

Lipids
(1) Phospholipids (75% of total lipids)
Composition : 1 glycerol, 2 fatty acids, and 1 phosphate group.
Amphipathic : phospholipids contain both polar (phosphate) and nonpolar (fatty acid) regions.
Phospholipid Bilayer : polar phosphate groups orient outward toward the water; nonpolar fatty acids
                          orient toward the middle of the double lipid layer.
(2) Cholesterol (20% of total lipids)   Strengthens the membrane and decreases its flexibility.
(3) Glycolipids (5% of total lipids)   Important for cell-to-cell communication and cohesion.

Proteins
(1) Integral Proteins   (often span the membrane; are an "integral" part of the membrane)
Amphipathic : integral proteins contain both polar and nonpolar regions (amino acid side chains).
Glycoproteins : most integral proteins are glycoproteins (sugar-protein combination).
Functions : channels (pores); transporters (carriers); receptors (recognition sites for ligands);
               enzymes; cell identity markers that allows cells to recognize foreign cells.
(2) Peripheral Proteins   (loosely attached to the inner and outer surfaces of the membrane)
Functions :   cytoskeleton anchors; cell shape; cell motility; act as enzymes.

MEMBRANE TRANSPORT
The plasma membrane regulates the movements of molecules and ions into and out of the cell.

Passive Processes   (processes that do not require expenditure of ATP)
(1) Simple Diffusion : the movement of molecules or ions due to their kinetic energy from a region of higher to lower concentration. Oxygen, carbon dioxide, fatty acids, and steroid hormones are examples of nonpolar molecules; since they are soluble in lipids, they diffuse rapidly through the lipid portions of the plasma membrane. Ions and small polar molecules (such as water) are insoluble in lipids; they diffuse through the water-filled channels formed by some of the integral proteins (ion channels).
(2) Osmosis : the diffusion of solvent (water) across a selectively permeable membrane. As in all kinds of diffusion, the water molecules move from a region of higher to lower concentration.
(3) Filtration : water and dissolved substances move across a membrane due to hydrostatic pressure.
(4) Facilitated Diffusion : lipid-insoluble molecules and molecules too large for the ion channels are carried through the membrane by transporter (carrier) proteins.

Active Processes   (processes that require ATP)
(1) Primary Active Transport : Energy of ATP directly moves a substance via *transporters*.
(2) Secondary Active Transport : Ion gradients drive substances across membranes. In *symports (cotransporters)* sodium ions and a second substance move in the same direction; in *antiports (contratransporters)* sodium ions and a second substance move in opposite directions.
(3) Endocytosis : in *phagocytosis* ("cell eating") pseudopods extend around a substance, enclose it, and bring it into the cell, forming a phagocytic vesicle; in *pinocytosis* ("cell drinking") the infolding of the plasma membrane traps droplets of extracellular fluid (ECF / fluid outside the cell), forming pinocytic vesicles; in *receptor-mediated endocytosis* substances bind to receptors on the plasma membrane, triggering an infolding of the membrane and the formation of an endocytic vesicle.
(4) Exocytosis : secretory vesicles fuse with the membrane and release their contents into the ECF.

MEMBRANE TRANSPORT
Fluid Mosaic Model

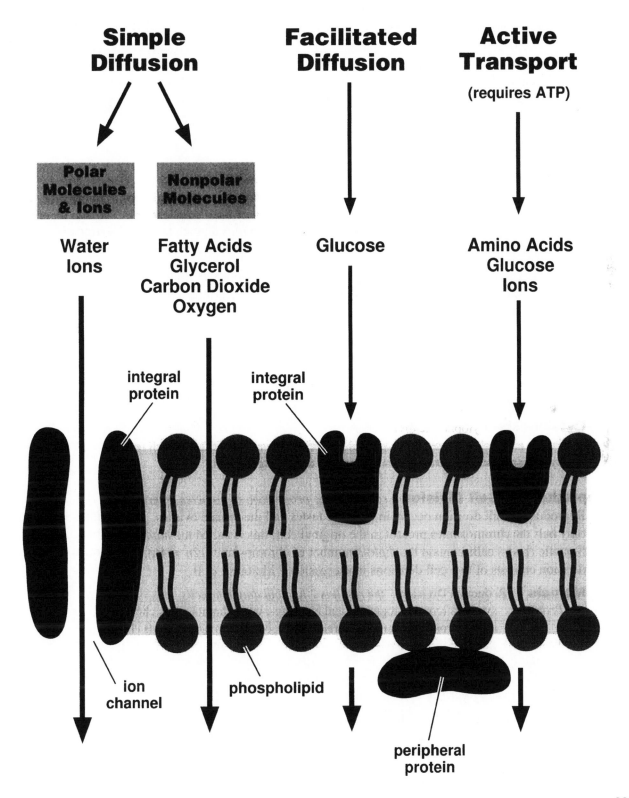

Simple Diffusion

Facilitated Diffusion

Active Transport (requires ATP)

Polar Molecules & Ions

Nonpolar Molecules

Water
Ions

Fatty Acids
Glycerol
Carbon Dioxide
Oxygen

Glucose

Amino Acids
Glucose
Ions

integral protein

integral protein

ion channel

phospholipid

peripheral protein

THE CELL / Nucleus

STRUCTURE

Nuclear Envelope (Nuclear Membrane) : a double membrane with pores that surrounds the nucleus.
Nuclear Pores : pores 10X as large as ion channels; allow passage of large molecules.
Nucleoli : one or more spherical bodies in the nucleus; where the subunits of ribosomes are assembled.
Chromatin : uncoiled DNA; it is present in nondividing cells as a fine network of threads coiled to
    various degrees (microscopically, it appears as dark regions on the periphery of the nucleus).
Chromosomes : condensed, highly coiled chromatin; form during cell division; consist of genes.
Gene : a unit of hereditary information; each gene is a portion of a DNA molecule.
Homologous Chromosomes : contain genes that control the same traits, but are not identical.
Chromatids : identical chromosomes attached by a centromere; the result of replication.

CELL DIVISION

Somatic Cell Division   *results in an increase in somatic (body) cells*

Interphase (metabolic phase)   When a cell is between divisions, it is said to be in interphase. During interphase each chromosome in the nucleus makes a copy of itself (replication). The two identical chromosomes that result from replication are called *chromatids* and are united by a structure called a *centromere*. Synthesis of substances needed for cell division also occurs at this time.

In somatic cell division a single starting cell (parent cell) duplicates itself to produce 2 identical cells (daughter cells). Two processes are involved : nuclear division (mitosis) and cytoplasmic division (cytokinesis).

Mitosis   (nuclear division) *the distribution of two sets of chromosomes into separate nuclei*
Prophase : chromosomes become visible; nucleus disappears; mitotic spindle appears.
Metaphase : chromosome pairs line up on the metaphase plate.
Anaphase : centromeres divide; identical chromosomes move to opposite poles.
Telophase : nucleus reappears; mitotic spindle disappears.

Cytokinesis (cytoplasmic division)
In late anaphase a furrow develops and progresses inward, dividing the cell into two separate portions of cytoplasm; the final result is 2 cells identical to the original cell.

Reproductive Cell Division   *results in the production of gametes (sperm or eggs)*

Reproductive cell division occurs in the male testes and the female ovaries. The daughter cells receive only half the chromosomes present in the original cell; this is called the *haploid* number ($n = 23$). Somatic (body) cells contain the *diploid* number of chromosomes ($2n = 46$). Reproductive cell division consists of two cell divisions in succession : Meiosis I & II.

Meiosis I (Reduction Division) *the result is 2 haploid daughter cells*
Prophase : synapsis (coming together) and crossing-over of homologous chromatids;
        during crossing-over genes are exchanged between homologous chromatids.
Metaphase : tetrads (homologous chromosome pairs) line up on the metaphase plate.
Anaphase : homologous chromosome pairs separate and move toward opposite poles.
Telophase : nuclear envelopes appear, forming two new nuclei.
Cytokinesis   The cytoplasm divides; 2 genetically different daughter cells result.

Meiosis II (Equatorial Division) *the 2 haploid daughter cells undergo mitosis*
There is no replication of chromosomes between the end of the 1st meiotic division and the start of the 2nd meiotic division; as a result, the daughter cells contain only half of the normal chromosome number. Prophase, metaphase, anaphase, and telophase are the same as they are in mitosis.
Cytokinesis   The cytoplasm divides; the final result is 4 genetically different gametes.

CELL DIVISION

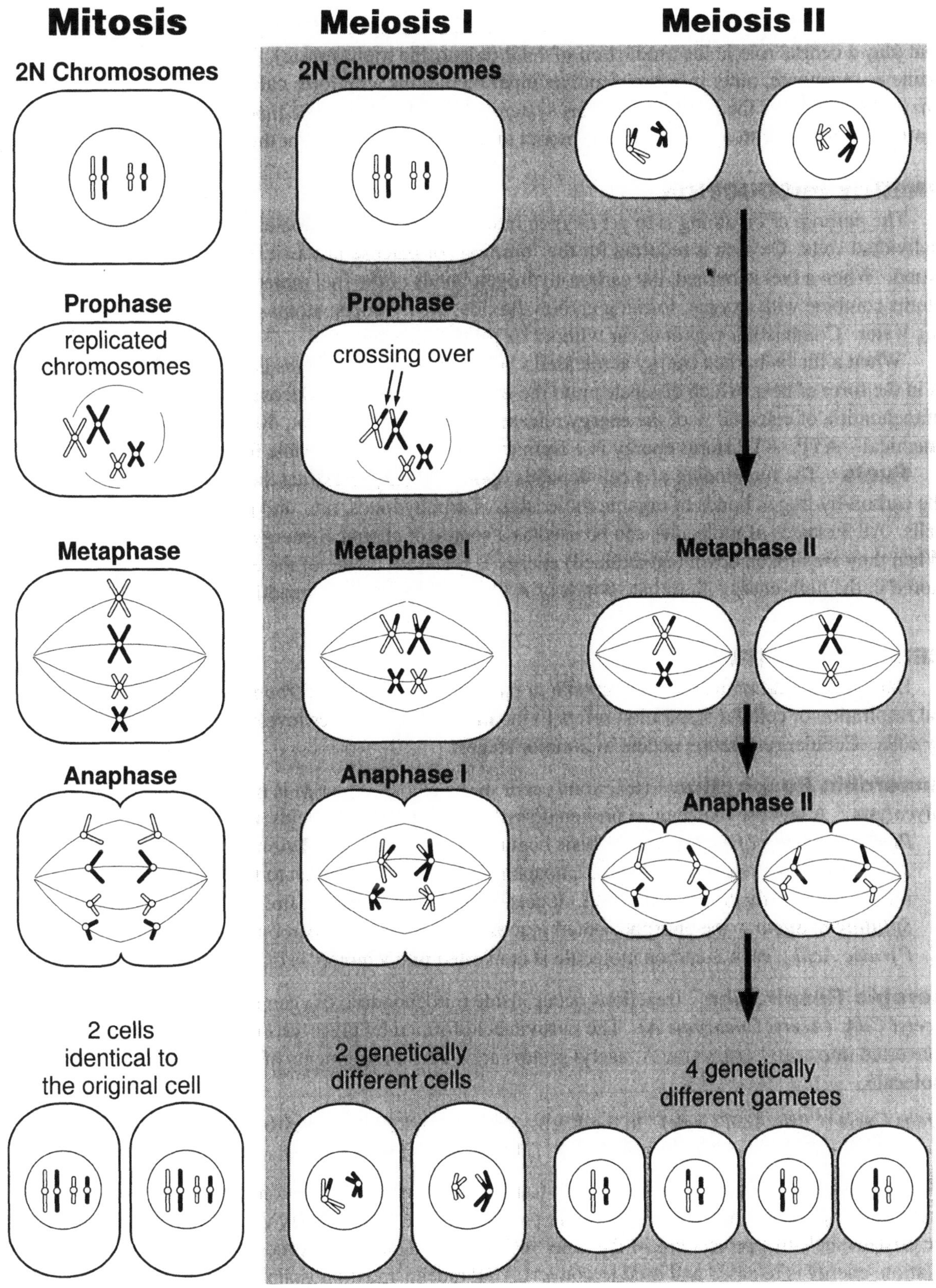

Mitosis

2N Chromosomes

Prophase

replicated
chromosomes

Metaphase

Anaphase

2 cells
identical to
the original cell

Meiosis I

2N Chromosomes

Prophase

crossing over

Metaphase I

Anaphase I

2 genetically
different cells

Meiosis II

Metaphase II

Anaphase II

4 genetically
different gametes

THE CELL / Mitochondria

STRUCTURE

Mitochondria (singular : mitochondrion) Mitochondria are oval-shaped or rod-shaped structures that play a central role in the production of ATP (adenosine triphosphate). They have a double membrane — a smooth, outer membrane and an inner membrane with folds called cristae, which contain the enzymes needed for the electron transport system. The space inside the inner membrane is called the matrix; it is filled with a fine granular material and contains enzymes for the Krebs cycle.

ENERGY PRODUCTION

The purpose of breathing is to get oxygen into the blood, and, ultimately, into the mitochondria of individual cells. Oxygen is required for the "burning" of glucose just as it is required for the burning of wood. When a fuel is burned, the carbon-hydrogen bonds of the fuel molecules are broken; the carbon atoms combine with oxygen, forming carbon dioxide; the hydrogen atoms combine with oxygen, forming water. Combustion cannot occur without oxygen.

When a fuel is burned energy is released. When wood or natural gas is burned, the energy released is in the form of heat, which dissipates into the environment. When glucose is burned (oxidized) in the mitochondria of cells, 60 % of the energy released is in the form of heat; 40 % is transferred to another chemical—ATP. ATP stores energy in a form that is immediately useable for cell functions.

Fuels The functioning of a cell depends upon its ability to extract the chemical energy locked in the carbon-hydrogen bonds of organic molecules. Carbohydrates, fats, and proteins serve as fuels for cells. All 3 classes of molecules can be used as a source of chemical energy for the synthesis of ATP. When they are broken down (catabolized) energy is released. Some of the energy released is trapped and stored in the high-energy phosphate bonds of ATP; the energy is then readily available for cell use.

CELLULAR RESPIRATION

External respiration refers to the uptake of oxygen and release of carbon dioxide by the lungs. Internal respiration or cellular respiration refers to the molecular processes involved in the production of ATP by cells. Cellular respiration occurs in 2 major stages:

Anaerobic Respiration (reactions occur in the cytosol; oxygen is not required)
Glycolysis : A glucose molecule is broken down, yielding 2 pyruvic acids and 2 ATP (net).
Phosphorylation of Sugar : Glycolysis begins with the phosphorylation of a glucose molecule; 2 ATP molecules are split, releasing 2 phosphate groups, which attach to the glucose molecule, causing it to reach its activation energy level. It acts like a match starting a fire.
Splitting of Sugar : the phosphorylated glucose splits into two 3-carbon molecules.
Pyruvic Acid : each 3-carbon molecule is converted into a molecule of pyruvic acid.

Aerobic Respiration (reactions occur inside mitochondria; oxygen is required)
Acetyl CoA (Acetyl Coenzyme A) The pyruvic acid formed by glycolysis enters a mitochondrion and is converted into acetyl coenzyme A; acetyl groups are 2-carbon fragments of the original 6-carbon glucose molecule.

Krebs Cycle (Citric Acid Cycle) In the Krebs cycle, acetyl groups (2-carbon fragments) are broken down, yielding energy-rich hydrogen atoms. These reactions occur in the matrix of mitochondria.

Electron Transport System In the electron transport system hydrogen atoms are separated into protons and energy-rich electrons. The electrons are passed along a chain of enzymes located on the surfaces of the cristae (shelf-like projections of the inner membrane). During this process, called oxidative phosphorylation, energy is released and used to convert ADP (adenosine diphosphate) molecules into ATP molecules.

42

CELLULAR RESPIRATION
Overview

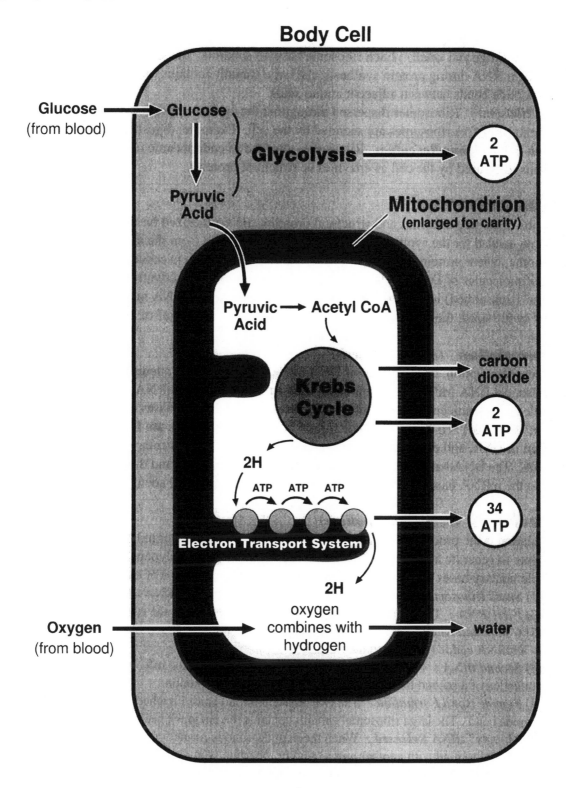

Body Cell

Glucose
(from blood)

Glucose

Glycolysis → 2 ATP

Pyruvic Acid

Mitochondrion
(enlarged for clarity)

Pyruvic → Acetyl CoA
Acid

Krebs Cycle → carbon dioxide

→ 2 ATP

2H

ATP   ATP   ATP

Electron Transport System → 34 ATP

2H
oxygen combines with hydrogen

Oxygen
(from blood)

water

Glucose + Oxygen + 2 ATP ⟶ Carbon Dioxide + Water + 38 ATP

THE CELL / Ribosomes

STRUCTURE

Ribosomal RNA and Ribosomal Proteins   Ribosomes are tiny spheres consisting of ribosomal RNA and ribosomal protein.

Subunits (large and small)   Each ribosome has two subunits: the small subunit attaches to a messenger RNA during protein synthesis; the large subunit contains the enzyme that is needed to form peptide bonds between adjacent amino acids.

Free Ribosomes   Ribosomes dispersed throughout the cytosol are called free ribosomes. Proteins synthesized on free ribosomes are secreted by the cell. Example: digestive enzymes.

Rough Endoplasmic Reticulum   Ribosomes attached to endoplasmic reticulum synthesize proteins to be used by the cell as enzymes or structural proteins.

PROTEIN SYNTHESIS

Proteins, including enzymes and structural proteins, are synthesized on ribosomes. The genetic information needed for the synthesis of a protein must be carried from the nucleus, where it is stored, to a ribosome, where protein synthesis occurs. Genetic information is coded in the nucleotide sequences of molecules of DNA. When a section of DNA (a gene) is activated, the genetic code is "rewritten" (transcribed) in the nucleotide sequence of a messenger RNA molecule; then, when a protein is synthesized, the code is "translated" using the new "language" of amino acid sequences.

Transcription   *(mRNA synthesis)*

During transcription, the sequence of DNA nucleotides is used as a template (pattern) for making a messenger RNA (mRNA). When being used as a template for mRNA synthesis, the double-stranded DNA molecule divides into single strands, exposing the bases of each strand. One strand (the sense strand) acts as the template for mRNA synthesis. There are four types of nucleotides present in DNA, and each type contains a base that has a complementary base in a nucleotide of mRNA. The DNA bases adenine (A), cytosine (C), guanine (G), and thymine (T) are complementary to the mRNA bases uracil (U), guanine (G), cytosine (C), and adenine (A), respectively.

Translation   *(polypeptide synthesis)*

Translation is the process by which the nucleotide sequence on a strand of mRNA is used as a template to generate a corresponding amino acid sequence in a polypeptide chain. The pairing of complementary bases plays a key role in translation. The sequence of events is the following:

(1) Small Ribosomal Subunit : A small subunit attaches to a mRNA and finds the *start codon.*

(2) First tRNA : The first tRNA carries a specific amino acid to the ribosome.

(3) Codon and Anticodon : The tRNA *anticodon* (triplet of nucleotides) attaches by base pairing to a mRNA *codon* (complementary triplet of nucleotides).

(4) Second tRNA : The ribosome moves 3 nucleotides along the mRNA to a second codon; the anticodon of a second tRNA attaches to the second mRNA codon.

(5) Peptide Bond Formation : The two adjacent amino acids attached to the tRNAs form a peptide bond. The large ribosomal subunit contains the enzymes needed for this reaction.

(6) "Empty" tRNA Released : When the peptide bond is formed, the first tRNA is released into the cytosol to bind with another amino acid of the same type.

(7) Synthesis Termination : The process continues until the ribosome reaches the *stop codon.*

PROTEIN SYNTHESIS

A messenger RNA carries genetic information coded in its
base sequences from the nucleus to a ribosome.

The messenger RNA acts as a template for building
a polypeptide (a specific sequence of amino acids).

Transfer RNAs deliver specific amino acids that match
the genetic codes present on the messenger RNA.

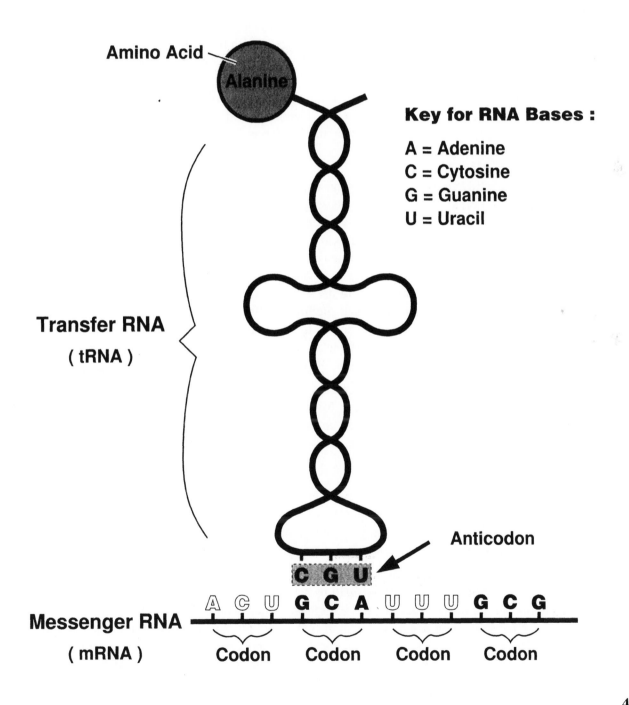

Amino Acid

Alanine

Key for RNA Bases :

A = Adenine
C = Cytosine
G = Guanine
U = Uracil

Transfer RNA

(tRNA)

Anticodon

C G U

A C U   G C A   U U U   G C G

Messenger RNA

(mRNA)   Codon   Codon   Codon   Codon

THE CELL / Cell Communication

LIGANDS and RECEPTORS

Ligands

A ligand is any ion or molecule that binds to a receptor protein by forces other than covalent bonds. All *chemical messengers* (hormones, neurotransmitters, paracrines, and autocrines) are ligands. Chemicals referred to as *second messengers*, such as calcium ions and cyclic AMP, are also ligands.

Protein Receptors

A protein receptor is a protein molecule that has an active site for a chemical messenger.

Active Site (also called *binding site*) The active site is a region of the protein that has a shape that is complementary to that of the chemical messenger; so the chemical messenger can fit into the active site. This property of an active site is referred to as *chemical specificity*. The active site also has charged and polarized regions that interact electrically with the ligand. The term *affinity* refers to how strongly a ligand is held to an active site, based on how well the complementary shapes fit together and how strong the electrical forces of attraction are.

Target Cell Recognition Some active sites are specific for only one type of chemical messenger. This selectivity allows a chemical messenger to "identify" a particular protein receptor on a particular type of cell; in this way a hormone can "recognize" its target cells.

RECEPTOR MODULATION

Allosteric Modulation

When a ligand binds to an active site it alters the distribution of forces within the protein receptor, and this affects its shape. If the protein receptor has a second active site, it may alter the affinity of that active site for a particular chemical messenger. Protein receptors of this type are called allosteric ("other shape") proteins.

Enzyme Activation If the allosteric protein is an enzyme, its activity can be altered in this way. The first active site is called the *regulatory site*. When a ligand (called a modulator molecule) binds to the regulatory site, it changes the shape of the second active site, which is called the *functional site*. If this increases the affinity of the functional site for its substrate, the enzyme has become activated; if it decreases the affinity of the functional site for its substrate, the enzyme has become inactivated. (A *substrate* is a molecule to which an enzyme attaches in an enzyme-catalyzed reaction.)

Covalent Modulation

The shape of a protein receptor is also changed by phosphorylation. When a phosphate group is attached to a protein receptor by a covalent bond, the negative charge of the phosphate group alters the distribution of forces within the protein, changing its shape.

Protein Kinases Protein kinases are enzymes that catalyze the transfer of phosphate from a molecule of ATP to a protein. This changes the shape of the protein, and therefore its function.

HORMONE RECEPTORS

Steroids

The receptors for steroid hormones are in the cytoplasm (cytosol) of the target cells. Because steroids are lipids, they pass easily through the plasma membranes and bind to their receptors. The receptor-hormone complex enters the nucleus and alters the activity of genes that regulate the synthesis of particular proteins (enzymes).

Peptides

The great majority of hormones are peptide hormones; they range in size from small peptides, consisting of 3 amino acids, to small proteins with over 200 amino acids. The receptors for peptide hormones are on the outer surface of target cell plasma membranes. The binding of hormone with receptor activates G–proteins located in the plasma membrane; each G–protein activates an enzyme called adenylate cyclase, which is located on the inner surface of the plasma membrane. Adenylate cyclase catalyzes the conversion of ATP to cyclic AMP. And cyclic AMP initiates a series of reactions that alter the activity of a specific set of proteins (enzymes).

LIGANDS AND RECEPTORS
Example : hormone – receptor interaction

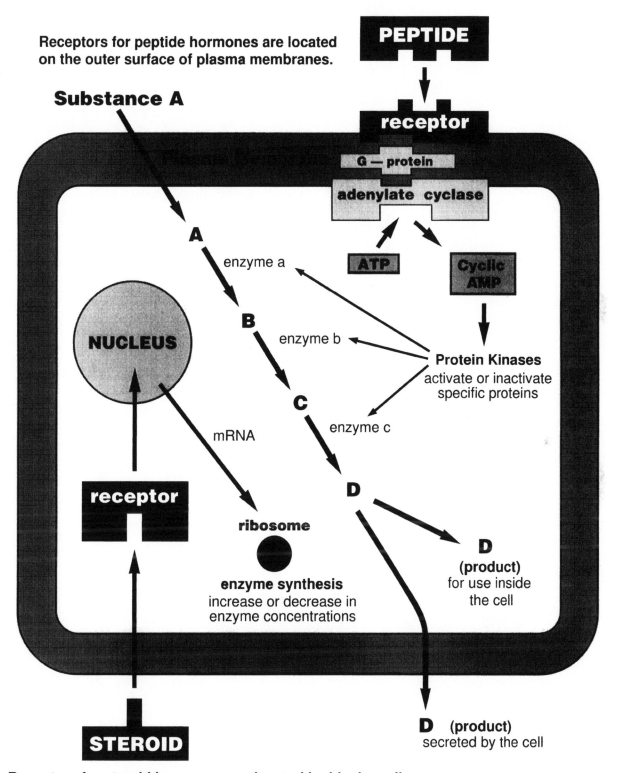

Receptors for peptide hormones are located on the outer surface of plasma membranes.

Substance A

PEPTIDE

receptor

G – protein

adenylate cyclase

ATP

Cyclic AMP

A

enzyme a

NUCLEUS

B

enzyme b

Protein Kinases
activate or inactivate
specific proteins

C

enzyme c

mRNA

D

receptor

ribosome

enzyme synthesis
increase or decrease in
enzyme concentrations

D
(product)
for use inside
the cell

D   (product)
secreted by the cell

STEROID

Receptors for steroid hormones are located inside the cell.

4 Tissues

Overview *50*
    1. Epithelial Tissues
    2. Connective Tissues
    3. Nervous Tissues
    4. Muscle Tissues
    5. Related Terms : Cell Junctions and Membranes

Epithelial Tissues *52*
    1. Covering and Lining Epithelium
        Simple : squamous, cuboidal, columnar (ciliated and nonciliated)
        Stratified : squamous, cuboidal, columnar, transitional
        Pseudostratified Columnar
    2. Glandular Epithelium
        Exocrine Glands
        Endocrine Glands

Connective Tissues *54*
    1. Loose Connective Tissue : areolar, adipose, reticular
    2. Dense Connective Tissue : regular, irregular, elastic
    3. Cartilage : fibrocartilage, hyaline, elastic
    4. Bone (Osseous) Tissue
    5. Blood (Vascular) Tissue

Nervous Tissues *56*
    1. Neurons   2. Neuroglia

Muscle Tissues *58*
    1. Skeletal Muscle Tissue   3. Smooth Muscle Tissue
    2. Cardiac Muscle Tissue

Integumentary System *60*
    1. Skin : epidermis and dermis
    2. Epidermal Derivatives : hairs, glands, and nails

Skin Functions *62*
    1. Sensation   5. Blood Reservoir
    2. Protection   6. Vitamin D Synthesis
    3. Immunity   7. Body Temperature Regulation
    4. Excretion

TISSUES / Overview

200 kinds of cells   About 200 distinct kinds of cells can be identified in the body.

4 broad categories of cells   All 200 kinds of cells can be classified in four broad categories
   based on structure and function: epithelial, connective, nervous, and muscle.

Primary Germ Layers   All tissues and organs of the body develop from 3 embryonic tissues:
   ectoderm, endoderm, and mesoderm.

EPITHELIAL TISSUE

Classification : (1) Covering & Lining Epithelium   (2) Glandular Epithelium

General Features

Cell Shapes : squamous (scale-like); cuboidal (cube-shaped); columnar (rectangular).

Apical Surface : exposed to a body cavity, lining of an internal organ, or the exterior.

Basal Surface : attached to the basement membrane.

Basement Membrane : attaches epithelium to underlying connective tissue.
   The basement membrane consists of two layers. The *basal lamina* is secreted by the epithelium;
   it contains collagen, laminin, and proteoglycans. The *reticular lamina* is deep to the basal lamina
   and is secreted by connective tissue; it contains reticular fibers, fibronectin, and glycoproteins.

Sheets : epithelial cells are arranged in continuous sheets.

Blood and Nerve Supply : nerves are present, but no blood vessels.

Functions : protection, filtration, lubrication, secretion, digestion, absorption, transportation, excretion, sensory reception, and reproduction.

CONNECTIVE TISSUE

Classification : (1) Loose   (2) Dense   (3) Cartilage   (4) Bone   (5) Blood

General Features

Matrix : consists of ground substance and fibers (collagen, elastic, and reticular).

Cell Types : fibroblasts (secrete matrix); macrophages (phagocytes); white blood cells;
   plasma cells (secrete antibodies); mast cells (produce histamine); adipocytes (store lipids).

Functions : binds together, supports, and strengthens other body tissues; protects and insulates internal organs; compartmentalizes structures such as skeletal muscles.

NERVOUS TISSUE

Classification : (1) Neurons   (2) Neuroglia

Functions : neurons conduct nerve impulses; neuroglia protect and support neurons.

MUSCLE TISSUE

Classification : (1) Skeletal   (2) Cardiac   (3) Smooth

Functions : provide motion, maintenance of posture, and heat production.

Related Terms

Cell Junctions

   (1) Tight Junctions : form fluid-tight seals between cells.

   (2) Anchoring Junctions : fasten cells to one another or to extracellular materials.

   (3) Communicating Junctions : form fluid-filled tunnels between cells.

Membranes   (epithelial layer overlying a connective tissue layer)

   (1) Mucous Membranes : line cavities that open to the exterior (lining of GI tract).

   (2) Serous Membranes : line body cavities (pericardial, pleural, and abdominopelvic cavities).

   (3) Cutaneous Membrane: the skin.

   (4) Synovial Membranes : line joints, bursae, and tendon sheaths (no epithelium present).

THE 4 BASIC TISSUES

Epithelial Tissue

Covering & Lining

squamous

cuboidal   **columnar**

Glandular

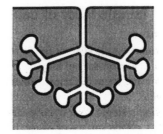

compound saccular glands
digestive glands (pancreas)
mammary glands

Connective Tissue

Loose Areolar

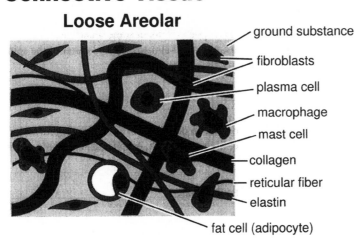

- ground substance
- fibroblasts
- plasma cell
- macrophage
- mast cell
- collagen
- reticular fiber
- elastin
- fat cell (adipocyte)

Compact Bone

lacuna (location of osteocyte)

matrix

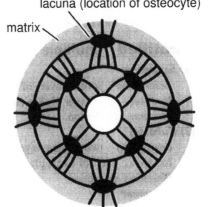

Nervous Tissue

neuron

Muscle Tissue

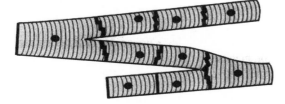

cardiac muscle

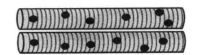
smooth muscle

skeletal muscle

TISSUES / Epithelial

COVERING and LINING EPITHELIUM

Simple   (single layer of cells)

(1) Squamous

   description : single layer of flat, scalelike cells.

   location : lines air sacs of lungs, glomerular (Bowman's) capsule of kidneys, and inner surface of eardrum.

   function : filtration, diffusion, osmosis, and secretion in serous membranes.

(2) Cuboidal

   description : single layer of cube-shaped cells.

   location : covers surface of ovary; lines anterior surface of lens; forms pigmented epithelium of retina;
         lines kidney tubules and smaller ducts of many glands.

   function : secretion and absorption.

(3) Nonciliated Columnar

   description : single layer of rectangular cells; some contain microvilli, some are goblet cells.

   location : lines gastrointestinal tract from stomach to anus, ducts of many glands, and gallbladder.

   function : secretion and absorption.

(4) Ciliated Columnar

   description : single layer of ciliated rectangular cells; contains goblet cells in some locations.

   location : lines upper respiratory tract, uterine tubes, uterus, paranasal sinuses, central canal of the spinal cord.

   function : moves fluids or particles along a passageway by ciliary action.

Stratified   (multiple layers of cells)

(1) Squamous

   description : several layers of cells; cuboidal to columnar shape in deep layers; squamous layers on top.

   location : outer layer of skin (keratinized); tongue surface; lining of mouth, esophagus, epiglottis, vagina.

   function : protection.

(2) Cuboidal

   description : two or more layers of cells; surface cells are cube-shaped.

   location : ducts of sweat glands; part of male urethra.

   function : protection.

(3) Columnar

   description : several layers of polyhedral cells; columnar cells only in the superficial layer.

   location : male urethra; large excretory ducts; portions of anal mucous membrane, and conjunctiva (eye).

   function : protection and secretion.

(4) Transitional

   description : variable appearance depending upon the degree of stretching (distention).

   location : lines urinary bladder and portions of ureters and urethra.

   function : permits distention.

Pseudostratified Columnar Epithelium   (not a true stratified tissue)

   description : all cells attached to basement membrane, but not all reach the surface; nuclei at different levels.

   location : upper respiratory tract; male urethra; auditory tubes; epididymis; lining of large glandular ducts.

   function : secretion and movement of mucus by ciliary action.

GLANDULAR EPITHELIUM

Exocrine Glands

(1) Salivary Glands   (3) Pancreatic Digestive Glands   (5) Ceruminous Glands (ear wax)

(2) Sweat Glands   (4) Sebaceous Glands (oil glands)

Endocrine Glands

(1) Anterior Pituitary   (4) GI Tract (enteroendocrine cells)   (7) Thymus

(2) Thyroid & Parathyroid   (5) Pancreas (pancreatic islets)   (8) Kidneys

(3) Ovaries & Placenta   (6) Testes & Prostate

EPITHELIAL TISSUES
Covering and Lining Epithelium

Simple

Stratified

Squamous
alveoli (air sacs)

skin (outer layer)

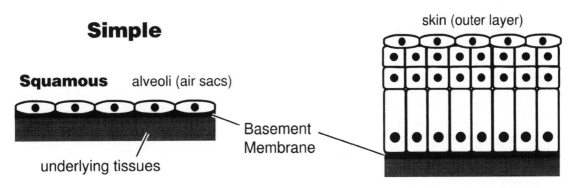

Basement
Membrane

underlying tissues

Cuboidal
kidney tubules

pharynx

male urethra

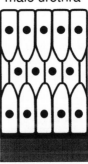

Columnar
G I tract

uterine tubules

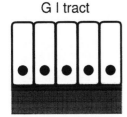

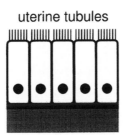

Pseudostratified Columnar
upper respiratory tract

Transitional
urinary bladder

stretched   contracted

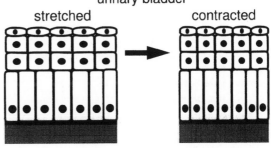

TISSUES / Connective

Loose Connective Tissue

(1) Areolar
composition : fibroblasts, macrophages, mast cells, adipocytes, and plasma cells;
  collagen, elastic, and reticular fibers ; semifluid ground substance.
location : skin (subcutaneous layer and papillary region); mucous membranes;
  around blood vessels, nerves, and body organs.
function : strength; elasticity; support.

(2) Adipose
composition : adipocytes; reticular fibers.
location : skin (subcutaneous layer); around heart and kidneys; yellow bone marrow;
  padding around joints; behind eyeball in eye socket.
function : energy reserve (stores triglycerides); insulation; support; protection.

(3) Reticular
composition : reticular cells; reticular fibers.
location : liver; spleen; lymph nodes; basal lamina (underlying epithelial cells).
function : binds together smooth muscle cells; forms framework (stroma) of organs.

Dense Connective Tissue

(1) Regular
composition : fibroblasts in rows between bundles of collagen fibers.
location : tendons (muscle to bone); ligaments (bone to bone); aponeuroses (muscle to muscle).
function : provides strong attachment between various structures.

(2) Irregular
composition : a few fibroblasts; predominantly collagen fibers; some elastic fibers.
location : skin (dermis); fasciae; periosteum of bone; perichondrium of cartilage; joint capsules;
  heart valves; membrane capsules around various organs (kidneys, liver, lymph nodes, testes).
function : provides strength.

(3) Elastic
composition : fibroblasts; elastic fibers.
location : lung tissue; elastic arteries; trachea and bronchial tubes; ligaments between vertebrae;
  true vocal cords; suspensory ligament of penis.
function : allows stretching.

Cartilage

(1) Fibrocartilage
composition : chondrocytes scattered among bundles of collagen fibers.
location : intervertebral discs; menisci of knee joint; pubic symphysis (anterior junction of hip bones).
function : support and fusion.

(2) Hyaline
composition : chondrocytes in a bluish white and glossy ground substance.
location : ends of long bones; anterior ends of ribs; nose; parts of larynx, trachea, and bronchial tubes.
function : allows movement at joints; flexibility; support.

(3) Elastic
composition : chondrocytes; elastic fibers.
location : external ear; auditory (Eustachian) tube; epiglottis of larynx.
function : gives support and maintains shape.

Bone (Osseous Tissue)

(1) Compact : consists of osteons that contain lamellae, lacunae, osteocytes, canaliculi, and central canals.
(2) Spongy : consists of a latticework of bony plates; spaces between the plates are filled with red bone marrow.

Blood (Vascular Tissue)

(1) Formed Elements : red blood cells, white blood cells, and platelets.
(2) Plasma : consists of water with dissolved nutrients, ions, hormones, gases, and plasma proteins.

CONNECTIVE TISSUES

Loose

Areolar

Adipose
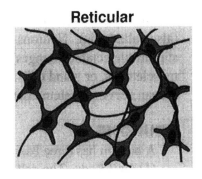

Reticular

Dense

Regular

Irregular
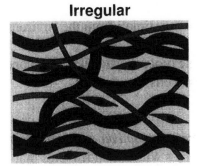

Elastic
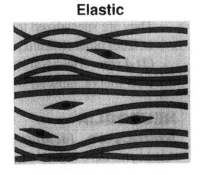

Cartilage

Fibrocartilage
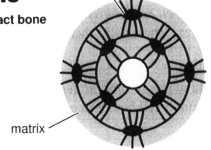

Hyaline
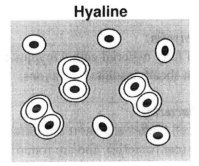

Elastic
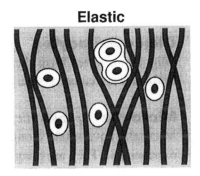

Bone

compact bone

lacuna
(location of osteocyte)

matrix

Blood

white blood cell

red blood cell

55

TISSUES / Nervous

Nervous tissue consists of 2 cell-type classifications: neurons (nerve cells) and neuroglia (also called glia). Neurons monitor changes in the internal and external environments, convert stimuli (detectable changes) into nerve impulses, and conduct these impulses to other neurons, muscle cells, or gland cells. Neuroglia insulate, nourish, support, and protect the neurons. The human nervous system contains at least 10 billion neurons and 100 - 500 billion neuroglial cells.

NEURONS
A neuron has three basic parts: the cell body, dendrites, and axon.

(1) Cell Body The cell body includes the nucleus and the cytoplasm immediately surrounding it. Its membrane is receptive to stimuli from other neurons. The cytoplasm in this region contains the normal cell organelles : mitochondria, ribosomes, Golgi complex, etc.

(2) Dendrite   Dendrites are highly branched processes that extend out from the cell body. They are specialized for receiving stimuli from the environment, specialized sensory organs, or from other neurons.

(3) Axon   The axon is a single process specialized for conducting nerve impulses to other cells (nerve, muscle, and gland cells).

NEUROGLIA
Astrocytes   *(neurotransmitter metabolism and potassium balance)*
Astrocytes have many long processes with expanded ends that attach to the walls of capillaries and form a sheath around them. They participate in the metabolism of neurotransmitters; maintain the proper balance of potassium for the generation of nerve impulses; help to form the blood-brain barrier that regulates the passage of substances into the brain.

Oligodendrocytes   *(form myelin)*
Oligodendrocytes form sheaths of fatty material called myelin around the axons of neurons in the brain and spinal cord. The myelin sheath insulates the axons.

Microglia   *(phagocytic; eat bacteria)*
Microglia have small, elongated cell bodies and short processes covered by numerous small expansions. They are phagocytic (eat bacteria), and are found in the brain and spinal cord.

Ependymal Cells   *(movement of cerebrospinal fluid)*
Ependymal cells line the ventricles of the brain and the central canal of the spinal cord. Most ependymal cells possess *cilia* that assist in the circulation of the cerebrospinal fluid.

Neurolemmocytes   *(form myelin sheath; support, protect, and nourish neurons)*
Neurolemmocytes form myelin sheaths around axons in nerves. The myelin sheath insulates the axons and increases the speed of impulse transmission.

Satellite Cells   *(support neuron cell bodies in ganglia of the peripheral nervous system)*
Satellite cells are flattened cells that surround and support neuron cell bodies in ganglia.

NERVOUS TISSUES
Neurons and Neuroglia

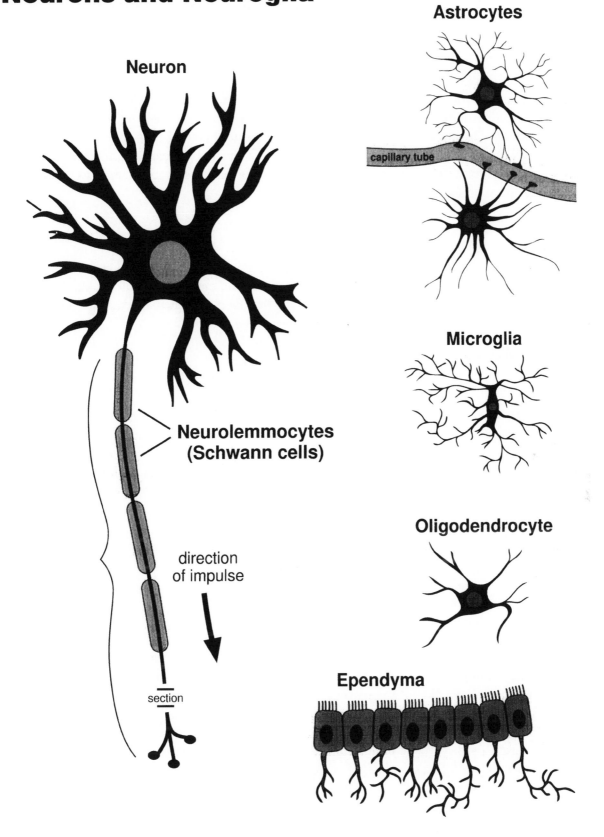

Neuron

Astrocytes

capillary tube

Neurolemmocytes (Schwann cells)

Microglia

direction of impulse

Oligodendrocyte

section

Ependyma

TISSUES / Muscle

There are three kinds of muscle tissue : skeletal, cardiac, and smooth. They differ from one another in their appearance under the microscope, their location in the body, and the mechanisms by which they are controlled by the nervous and endocrine systems.

Muscle tissue has several main functions : motion, the stabilization of body positions (posture), the regulation of organ volume, and the generation of heat (thermogenesis).

Muscle tissue has four principal characteristics : excitability (irritability), contractility, extensibility, and elasticity. Excitability is the ability of a muscle cell to respond to neurotransmitters or hormones by producing electrical signals called action potentials (impulses). Contractility is the ability of a muscle cell to shorten (contract). Extensibility means that muscle tissue can be stretched without damaging effects. Elasticity means that muscle tissue tends to return to its original shape after shortening or stretching.

Skeletal Muscle Tissue
Description   Skeletal muscles are so-named because they are usually attached to bones. Skeletal muscle tissue consists of cylindrical fibers (cells) which contain many peripheral nuclei. Under the microscope light and dark bands are visible, giving the fibers a striated appearance. These muscles are under voluntary control by the somatic nervous system.
Location   Usually attached to bones.
Function   Motion, posture, and heat production (thermogenesis).

Cardiac Muscle Tissue
Description   Cardiac muscle is so-named because it forms most of the structures of the heart. It consists of branched, cylindrical, striated fibers (cells) which contain one or two centrally located nuclei. The cells are attached by thickened regions of plasma membrane (intercalated discs). Contraction of cardiac muscle is under involuntary control by the autonomic nervous system and hormones. Cardiac muscle is also called myocardium.
Location   The heart.
Function   The function of cardiac muscle tissue is to pump blood to all parts of the body.

Smooth Muscle Tissue
Description   Smooth muscle is so-named because it has a smooth (nonstriated) appearance under the microscope. It consists of spindle-shaped, nonstriated fibers (cells) which contain one centrally located nucleus. Smooth muscle is usually under involuntary control by the autonomic nervous system and hormones. There are two kinds of smooth muscle tissue : visceral (single-unit) and multiunit.
Location   Visceral (single-unit) smooth muscle tissue is found in the walls of the small arteries, veins, and hollow organs (stomach, intestines, uterus, urinary bladder, and gallbladder). Multiunit smooth muscle tissue is found in the walls of large arteries, airways to the lungs, arrector pili muscles attached to hair follices, and radial and circular muscles of the iris that adjust pupil size.
Functions   Functions include the constriction of blood vessels and airways, the propulsion of foods through the gastrointestinal tract, and contraction of the urinary bladder and gallbladder.

MUSCLE TISSUES

Skeletal Muscle

Skeletal Muscle Cell (longitudinal section)

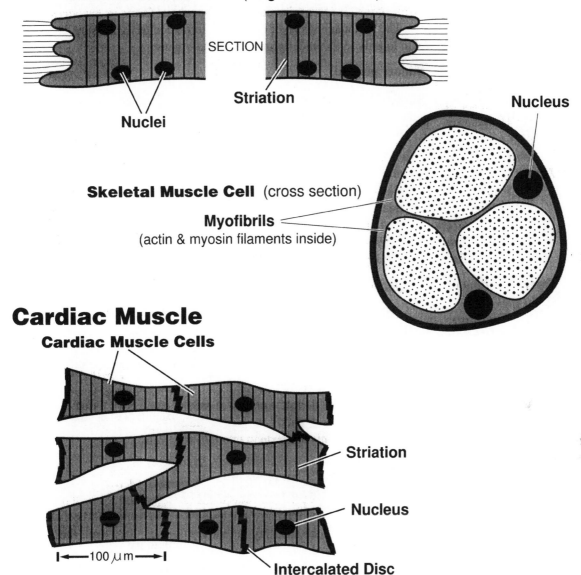

SECTION

Nuclei

Striation

Nucleus

Skeletal Muscle Cell (cross section)

Myofibrils
(actin & myosin filaments inside)

Cardiac Muscle

Cardiac Muscle Cells

Striation

Nucleus

|←—100 μm—→|

Intercalated Disc

Smooth Muscle

Smooth Muscle Cells

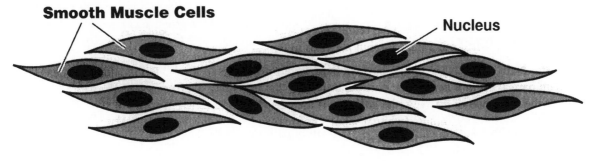

Nucleus

TISSUES / Integumentary System

SKIN

The skin has two principal parts : the outer portion is called the epidermis and is composed of epithelial tissue; the inner portion is called the dermis and is composed of connective tissue.

Beneath the dermis is the *subcutaneous layer* (also called the *superficial fascia* or *hypodermis*), which is not considered part of the skin. The subcutaneous layer consists of loose connective tissues (adipose and areolar tissues); it binds the skin loosely to the underlying tissues and organs.

Epidermis   (outer, thinner portion)
Cell Types
   (1) Keratinocytes : produce keratin (waterproof the skin).
   (2) Melanocytes : produce melanin (absorb UV light).
   (3) Langerhans cells : involved in the immune response; phagocytize antigens (foreign molecules).
   (4) Tactile Discs (Merkel's discs) : sensory receptors for touch.
Layers   (arranged from superficial to deep)
   (1) Stratum Corneum : flat, dead cells completely filled with keratin.
   (2) Stratum Lucidum : clear, flat, dead cells; present in thick skin of palms and soles.
   (3) Stratum Granulosum : flattened cells; produce the precursor of keratin.
   (4) Stratum Spinosum : polyhedral cells; keratinocytes take in melanin by phagocytosis.
   (5) Stratum Basale : single layer of cuboidal to columnar cells; stem cells produce keratinocytes.

Dermis   (inner, thicker portion)
Cell Types
   (1) Fibroblasts : produce fibers (collagen, reticular, and elastic) and ground substance.
   (2) Macrophages : phagocytize bacteria and foreign substances.
   (3) Adipocytes (fat cells) : store triglycerides (energy reservoir); insulate; cushion.
   (4) Sensory Receptors : free and encapsulated nerve endings sensitive to specific stimuli.
Regions
   (1) Papillary Region : outer 1/5 of the dermis; consists of areolar connective tissue.
   (2) Reticular Region : inner 4/5 of the dermis; consists of dense, irregular connective tissue.
   Dermal Papillae : projections of dermis that extend into the epidermis.

EPIDERMAL DERIVATIVES

Hairs (Pili) Hair is composed of dead, keratinized cells welded together. A hair consists of a *shaft* above the surface, a *root* that penetrates the dermis and subcutaneous layer, and a *hair follice* that surrounds the root. *Arrector pili muscles* attached to hair follicles contract in response to cold, causing goose pimples. Hair on the head protects against the sun's rays and decreases heat loss; eyelashes, eyebrows, and hair in the nostrils and external ear canals block the entrance of foreign particles.

Glands Several kinds of glands are associated with the skin. *Sebaceous glands* secrete an oily substance called *sebum* and are usually associated with hair follicles; sebum moistens hairs and waterproofs the skin. There are two principal types of *sudoriferous (sweat) glands* : *eccrine* sweat glands are most numerous in the palms and the soles; *appocrine* sweat glands are found mainly in the skin of the axilla (armpit), pubic region, and areolae (pigmented regions) of the breasts. *Ceruminous glands* secrete a waxy substance which combines with the secretions of sebaceous glands to produce *cerumen*. Cerumen provides a sticky barrier that traps foreign particles in the ear canal.

Nails Nails are plates of tightly packed, hard, keratinized cells of the epidermis. Each nail has three main parts : the *nail body* is the portion of the nail that is visible; the *free edge* is the part that may extend past the distal end of the digit; the *nail root* is the portion that is buried in a fold of skin. Other nail structures include the *lunula* (whitish semilunar area) and the *eponychium* (cuticle). The *nail matrix* is a region of epithelium under the nail root; growth of nails occurs by the transformation of superficial cells of the matrix into nail cells.

SKIN

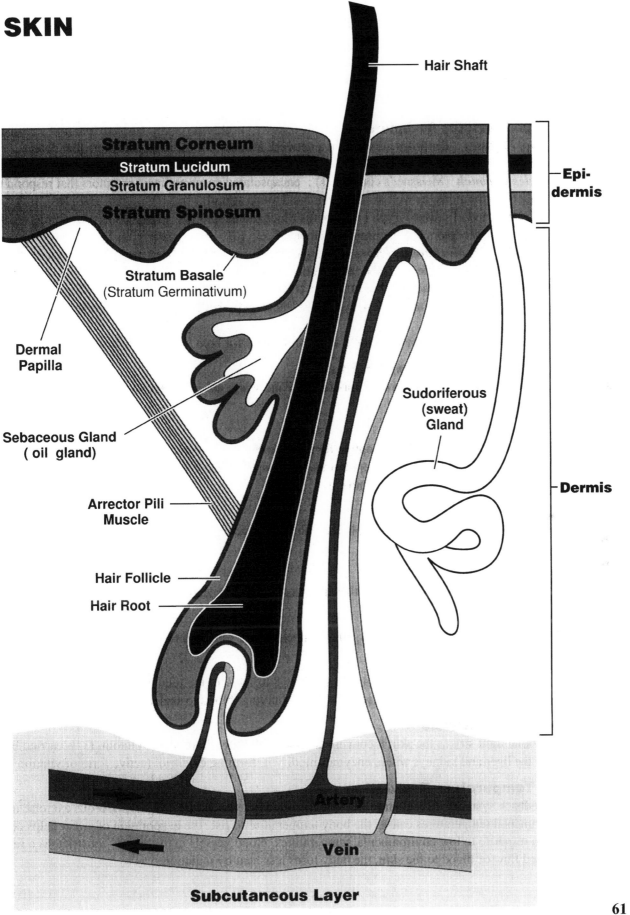

Hair Shaft

Stratum Corneum
Stratum Lucidum
Stratum Granulosum
Stratum Spinosum

Epi-
dermis

Stratum Basale
(Stratum Germinativum)

Dermal
Papilla

Sudoriferous
(sweat)
Gland

Sebaceous Gland
(oil gland)

Dermis

Arrector Pili
Muscle

Hair Follicle

Hair Root

Artery

Vein

Subcutaneous Layer

61

SENSORY SYSTEM / Skin Functions

(1) Sensation
Receptors in the skin monitor 3 basic types of cutaneous sensations : tactile, thermal, and pain.

Touch and Pressure Receptors
Hair Root Plexuses : dendrites arranged in a network around hair follicles; receptors that detect movement when hairs are disturbed.

Corpuscles of Touch (Meissner's corpuscles) : encapsulated nerve endings; receptors that respond to low frequency vibrations, as well as to pressure and touch stimuli.

Tactile Discs (Merkel's discs / Type I Cutaneous Mechanoreceptors) : expanded nerve endings (flattened dendrites); receptors for discriminative touch.

Type II Cutaneous Mechanoreceptors (end organs of Ruffini) : expanded nerve endings embedded in the dermis; receptors that detect heavy and continuous touch.

Lamellated Corpuscles (Pacinian corpuscles) : oval structures composed of a connective tissue capsule, layered like an onion, that enclose a dendrite; receptors that respond to pressure and high frequency vibrations.

Free Nerve Endings Free nerve endings are the receptors for both tickle and itch sensations.

Thermal Receptors *(Thermoreceptors)*
Free Nerve Endings The sense receptors for cold and warm are called thermoreceptors. They are free (naked) nerve endings.

Pain Receptors *(Nociceptors)*
Free Nerve Endings The sense receptors for pain are called nociceptors. They are free (naked) nerve endings located between cells of the epidermis. Nociceptors respond to all types of high intensity stimuli and stimuli that cause tissue damage.

(2) Protection
The skin protects underlying tissues from physical abrasion, bacterial invasion, dehydration, and ultraviolet radiation.

(3) Immunity
Langerhans cells of the epidermis phagocytize antigens (foreign molecules).

(4) Excretion
A small amount of salts and and several organic compounds are removed from the body in sweat.

(5) Blood Reservoir
Extensive networks of blood vessels in the dermis carry 8 to 10 % of the total blood flow in a resting adult. During vigorous exercise skin blood vessels constrict, shunting blood to skeletal muscles.

(6) Vitamin D Synthesis
When ultraviolet light strikes the skin it converts a provitamin into vitamin D_3. Vitamin D_3 is carried by the blood to the liver and kidneys where enzymes modify it, forming calcitriol (active form of vitamin D).

(7) Body Temperature Regulation
Negative feedback systems prevent large fluctuations in body temperatures. When vigorous exercise or high environmental temperatures causes the body temperature to rise, the evaporation of sweat helps cool the body. In response to low environmental temperatures, blood vessels in the dermis constrict; as a result of the reduced flow of blood to the skin, the body loses less heat by radiation.

SKIN RECEPTORS
Touch, Temperature, Pain, and Pressure

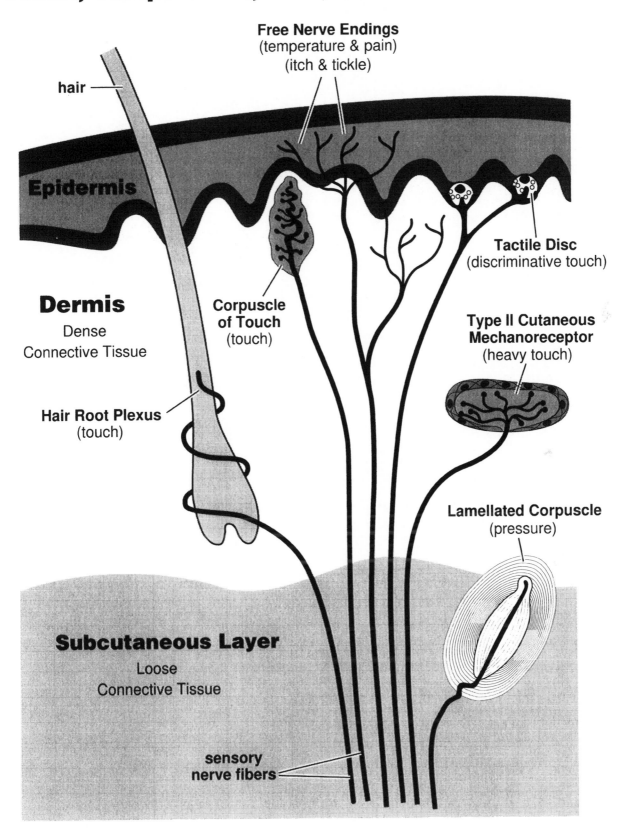

Free Nerve Endings
(temperature & pain)
(itch & tickle)

hair

Epidermis

Tactile Disc
(discriminative touch)

Dermis
Dense
Connective Tissue

**Corpuscle
of Touch**
(touch)

**Type II Cutaneous
Mechanoreceptor**
(heavy touch)

Hair Root Plexus
(touch)

Lamellated Corpuscle
(pressure)

Subcutaneous Layer
Loose
Connective Tissue

sensory
nerve fibers

5 Homeostasis

Survival : The Unifying Theme *66*
1. Survival of an Organism
    The Basic Necessities
2. Survival of Individual Cells
    Single-Celled Organisms
    Multicellular Organisms
3. Survival of a Species

The Internal Environment *68*
1. Extracellular Fluid
2. Single-Celled vs. Multicellular Organisms
3. Homeostasis
4. Fluid Compartments
    ICF (Intracellular Fluid) : the fluid inside cells
    ECF (Extracellular Fluid) : interstitial fluid and plasma

Feedback Systems *70*
1. Homeostatic Mechanisms
    Negative Feedback
    The Reflex Arc : receptor, control center, and effector
2. Factors Regulated
    Chemical : nutrients, gases, electrolytes, and water
    Physical : temperature, pressure, and volume
3. Examples of Feedback Systems
    Blood Pressure
    Blood Sugar
    Blood pH

HOMEOSTASIS / Survival : The Unifying Theme

The structures and functions of the body have one purpose in common — survival.
Survival of cells, survival of the individual, and survival of the species.

SURVIVAL OF AN ORGANISM
The Basic Necessities
> (1) Food, water, and oxygen
> (2) Protection from predators and parasites
> (3) Shelter from environmental extremes

SURVIVAL OF INDIVIDUAL CELLS
Single-Celled Organisms
(a watery environment that is naturally balanced & stable)
All single-celled organisms are aquatic; they live in lakes, rivers, and oceans. The relatively
large volume of water that surrounds them prevents sudden and extreme environmental
changes; it is well known that ocean temperatures change very slowly, while a puddle of
water can drop to freezing temperatures overnight. Besides offering a stable environment,
the water contains abundant supplies of nutrient molecules and dissolved oxygen. The
metabolic wastes (poisonous in concentrated amounts) produced by these cells diffuses away.

Multicellular Organisms
(a watery environment that is constantly changing)
The cells of multicellular organisms are also surrounded by a watery fluid, which is called
the extracellular fluid (*extra* = outside; outside the cell). To stay alive every cell in the body
must have a thin covering of this fluid; the exchange of nutrients and gases with the blood
can occur only through a liquid medium. Unlike the environments of single-celled organisms, the volume of the extracellular fluid is extremely small in relation to the cells, so it is
very unstable. An active cell in a multicellular organsim radically alters its environment in
an instant: the oxygen and nutrients of the surrounding fluid are quickly depleted; poisonous
ammonia is released; carbon dioxide is released, which makes the environment more acidic.

> ***Organ Systems***   Coordinated efforts by the organ systems of the body keep the contents
> of the extracellular fluid relatively constant. This is the common function of all the
> systems of the body, except for the immune and reproductive systems. The immune
> system protects the cells from pathogenic organisms (i.e., bacteria, viruses, fungi, para-
> sitic worms).

SURVIVAL OF A SPECIES
The reproductive system is responsible for the survival of the species: the fusion of egg and
sperm produce a new individual; the recombination of genetic material during this fusion results
in genetic variation within the species, which improves the chances of species survival over long
periods of gradual environmental changes.

SURVIVAL OF INDIVIDUAL CELLS

Single-Celled Organisms
a water environment that is naturally balanced & stable

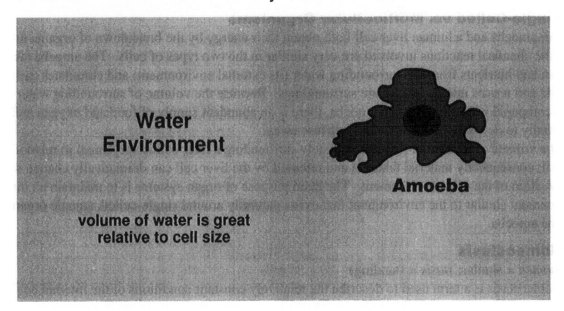

Water Environment

Amoeba

volume of water is great
relative to cell size

Multicellular Organisms
a watery environment that is constantly changing

Tissue Cells

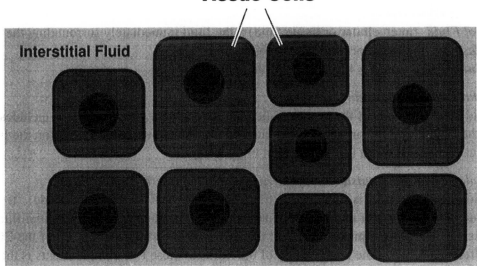

Interstitial Fluid

HOMEOSTASIS / The Internal Environment

Extracellular Fluid (ECF)
The internal environment is another name for the ECF—the blood plasma and the interstitial fluid. It is vital that the concentrations of substances and the pressure and temperature of the ECF remain relatively constant.

Single-Celled vs. Multicellular Organisms
An amoeba and a human liver cell both obtain their energy by the breakdown of organic nutrients; the chemical reactions involved are very similar in the two types of cells. The amoeba takes in oxygen and nutrients from the surrounding water (its external environment) and eliminates carbon dioxide and wastes into its immediate surroundings. Because the volume of surrounding water is great compared with the size of an amoeba, there is an abundant supply of food and oxygen and potentially toxic waste products quickly diffuse away.

The volume of interstitial fluid immediately surrounding a liver cell is very small in relation to the cell; consequently material taken in and released by the liver cell can dramatically change the composition of the cell's environment. The main purpose of organ systems is to maintain an internal environment similar to the environment that exists naturally around single-celled, aquatic organisms like the amoeba.

Homeostasis
(*homeo* = similar; *stasis* = standing)
Homeostasis is a term used to describe the relatively constant conditions of the internal environment. In order for an organism to function optimally, its individual cells must be surrounded by a fluid of closely regulated composition. Chemical factors such as blood concentrations of nutrients, gases, electrolytes, and water must be maintained at relatively constant levels. Also, physical factors such as body temperature, blood pressure, and blood volume must be maintained within narrow limits. This concept of an internal environment and the necessity of maintaining a relatively constant composition is the single most important unifying idea in the study of human physiology.

FLUID COMPARTMENTS
3 Fluid Compartments:
> *Interstitial Fluid* (intercellular or tissue fluid) : the fluid immediately surrounding the cells.
> *Blood Plasma :* the liquid portion of the blood.
> *Intracellular Fluid* (*intra* = inside) : the fluid inside the cells.

ICF *Intracellular Fluid (fluid inside the cells)*
The total fluid inside the cells of the body is called the intracellular fluid or ICF. It includes the cytosol and the fluid inside the organelles and the nucleus. In an average sized person the ICF contains about 28 liters of water (2/3 of the total body water).

ECF *Extracellular Fluid (fluid outside the cells; also called the internal environment)*
The total fluid outside the cells is called the extracellular fluid or ECF (*extra* = outside). It contains about 14 liters of water and is subdivided into 2 compartments : the interstitial fluid and the blood plasma. The plasma makes up 1/4 of the ECF; the interstitial fluid is 3/4 of the ECF. The main difference between the composition of the fluids in these two compartments is the presence of plasma proteins (fibrinogen, albumin, and globulins) in the plasma. Plasma is 7% protein molecules, while interstitial fluid has virtually no protein.

FLUID COMPARTMENTS

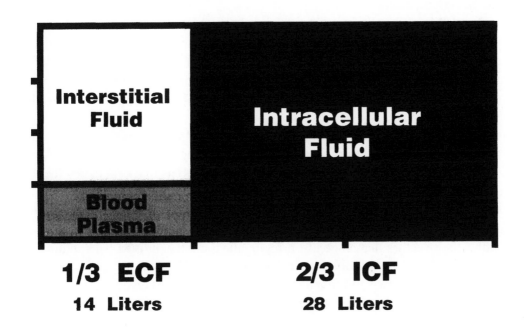

Interstitial Fluid	**Intracellular Fluid**
Blood Plasma	

1/3 ECF — 14 Liters   **2/3 ICF** — 28 Liters

Capillary Bed

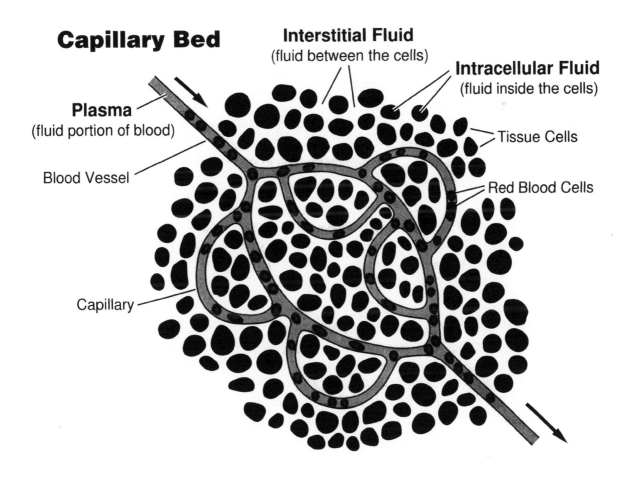

Interstitial Fluid
(fluid between the cells)

Intracellular Fluid
(fluid inside the cells)

Plasma
(fluid portion of blood)

Tissue Cells

Blood Vessel

Red Blood Cells

Capillary

HOMEOSTASIS / Feedback Systems

Homeostatic Mechanisms

Information about all important aspects (chemical and physical) of the external and internal environments must be monitored continuously. On the basis of this information, "instructions" are sent to the various tissue cells (particularly muscle and gland cells) directing them to increase or decrease their activities. This processing of information is performed primarily by the nervous and endocrine systems. These two systems contribute to the survival of the organism by controlling and coordinating the body functions.

Negative Feedback Any change in the body's internal environment automatically initiates a chain of events that oppose it; this type of response is called negative feedback.

The Reflex Arc The reflex arc summarizes the sequence of events that occurs in a feedback system (during a homeostatic response). A reflex arc has three main components:

(1) receptor : the structure that is sensitive to a particular type of change in the environment.

(2) control center : the structure that interprets the change and decides how to respond.

(3) effector : the structure that responds, bringing the altered condition back to normal. Effectors are usually muscles or glands.

In nervous system responses two additional components are often involved :

The *afferent pathway* carries information from receptors to the brain or spinal cord via nerves.

The *efferent pathway* carries instructions from the brain or spinal cord to effectors via nerves.

Factors Regulated

Chemical : nutrients, gases, electrolytes, and water.

Physical : body temperature, blood pressure, and blood volume.

Examples of Feedback Systems

Blood Pressure

When the blood pressure drops below normal, pressure receptors in the major arteries respond to the change by sending nerve impulses to the brain. Centers in the brain send nerve impulses to the heart, increasing the rate and force of contractions, and to the blood vessels, causing constriction; blood pressure returns to normal.

Blood Sugar

When blood sugar levels rise too far above normal, beta cells in the pancreas detect the change and release the hormone insulin into the blood. Insulin increases the uptake of glucose by all cells of the body and increases the rate at which glucose is stored in liver cells as glycogen; the effect is to lower the blood glucose concentration.

Blood pH

Increased muscular activity during exercise increases the rate of cellular respiration, releasing large amounts of carbon dioxide into the blood. The carbon dioxide combines with water, forming carbonic acid, increasing the acidity (lowers the pH). Nerve cells in the respiratory center of the brain respond to increased hydrogen ion concentrations by sending nerve impulses to the respiratory muscles. As the depth and rate of breathing increases, more carbon dioxide is excreted by the lungs. As the blood concentration of carbon dioxide decreases, carbonic acid is converted back into carbon dioxide and water, lowering the hydrogen ion concentration.

FEEDBACK SYSTEMS (Reflex Arcs)

A feedback system helps to maintain a stable internal environment.

The 3 basic components of a feedback system are :
receptor, control center, and effector.

Thermostat

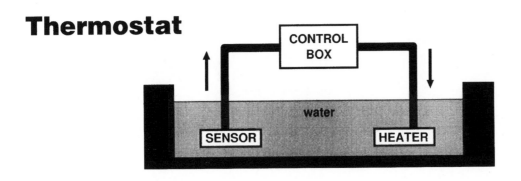

Feedback System
(nervous system responses)

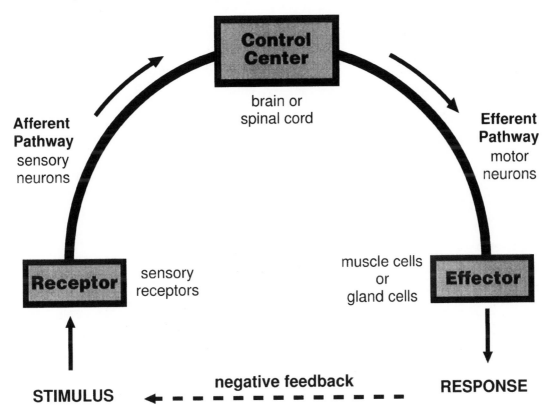

Part II : Self-Testing Exercises

Unlabeled illustrations from Part I

Chapter 1 : Structural Organization   74

Chapter 2 : Chemistry   81

Chapter 3 : The Cell   90

Chapter 4 : Tissues   96

Chapter 5 : Homeostasis   103

LEVELS OF ORGANIZATION

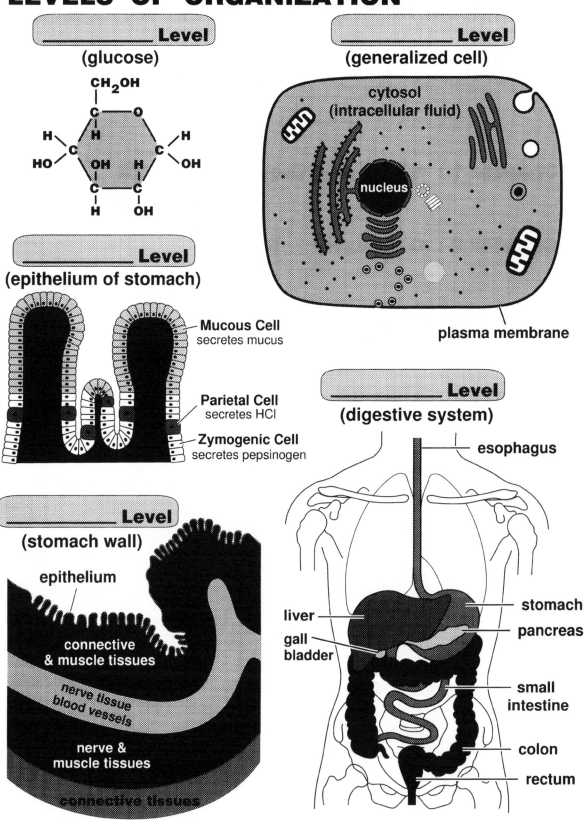

_____ Level
(glucose)

_____ Level
(generalized cell)

cytosol
(intracellular fluid)

nucleus

plasma membrane

_____ Level
(epithelium of stomach)

Mucous Cell
secretes mucus

Parietal Cell
secretes HCl

Zymogenic Cell
secretes pepsinogen

_____ Level
(digestive system)

esophagus

liver

gall
bladder

stomach

pancreas

small
intestine

colon

rectum

_____ Level
(stomach wall)

epithelium

connective
& muscle tissues

nerve tissue
blood vessels

nerve &
muscle tissues

connective tissues

SKELETAL SYSTEM
an example of an organ system

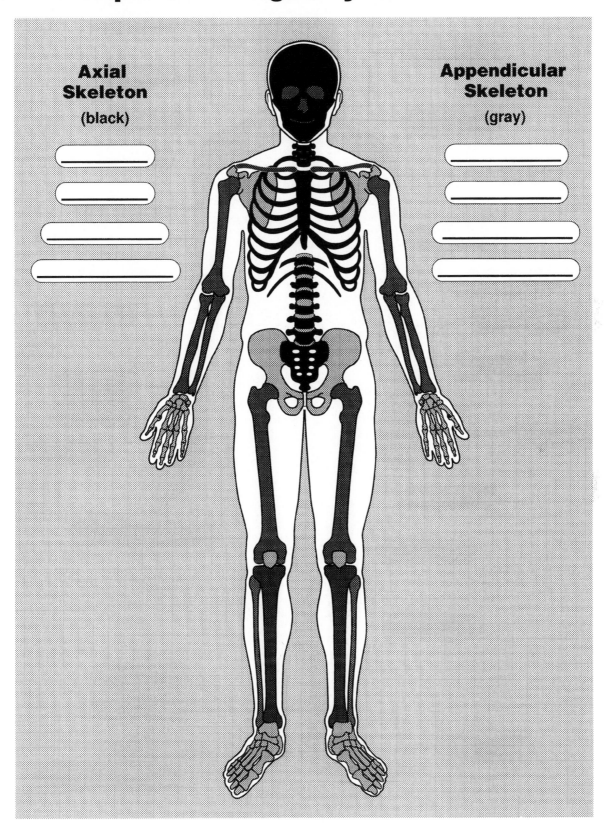

Axial Skeleton

(black)

Appendicular Skeleton

(gray)

REGIONAL NAMES

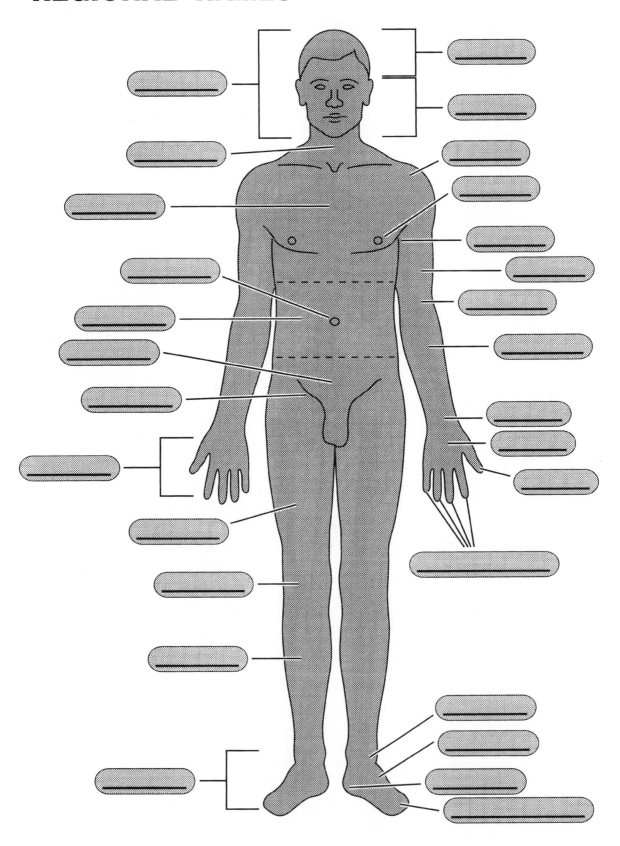

PLANES AND DIRECTIONAL TERMS

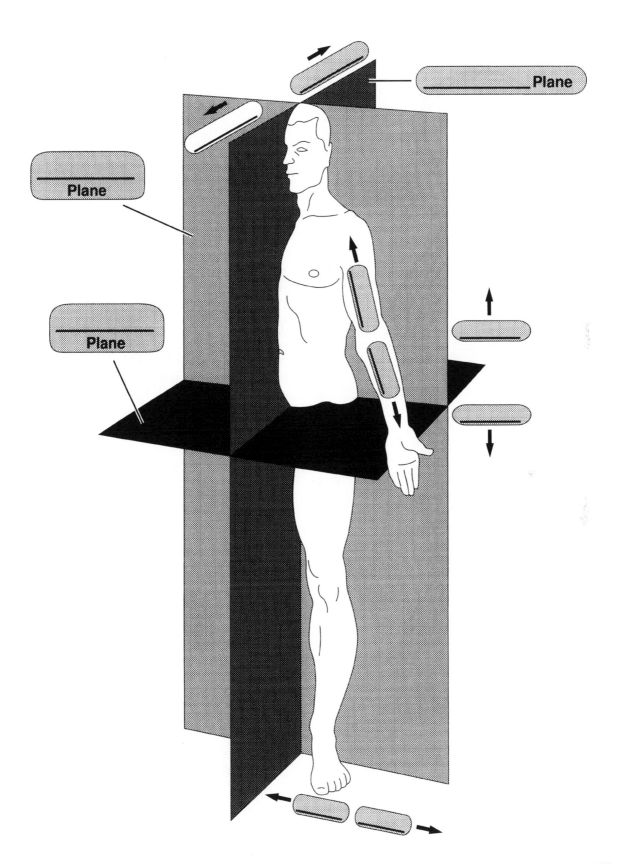

_____ Plane

_____ Plane

_____ Plane

BODY CAVITIES AND REGIONS

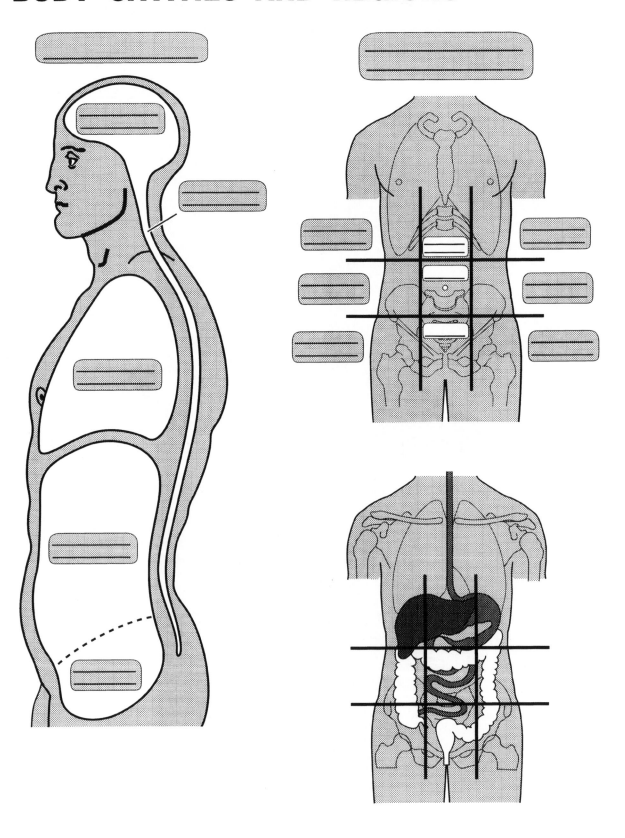

MEDIASTINUM

Transverse section at the level of the 6th thoracic vertebra.

The mediastinum includes all the contents of the thoracic cavity *except* the lungs.

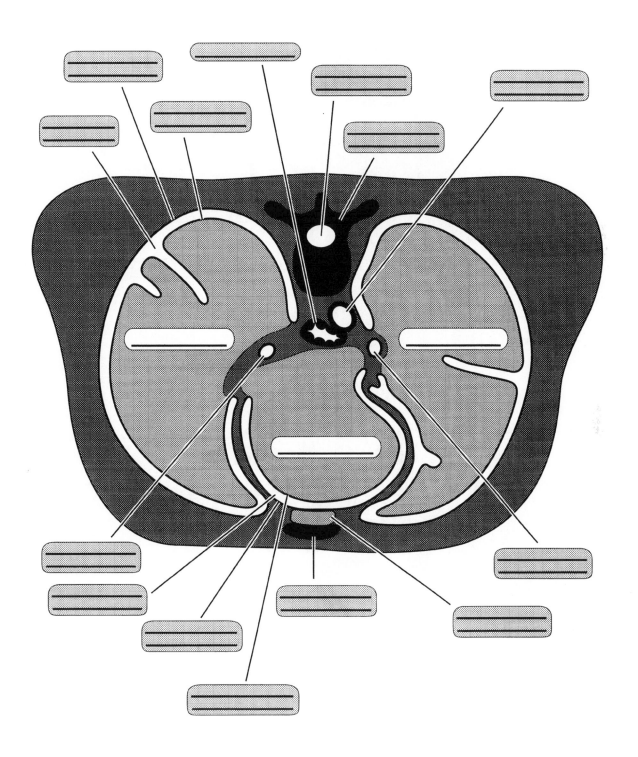

ABDOMINOPELVIC CAVITY
showing peritoneum and viscera

Midsagittal Section

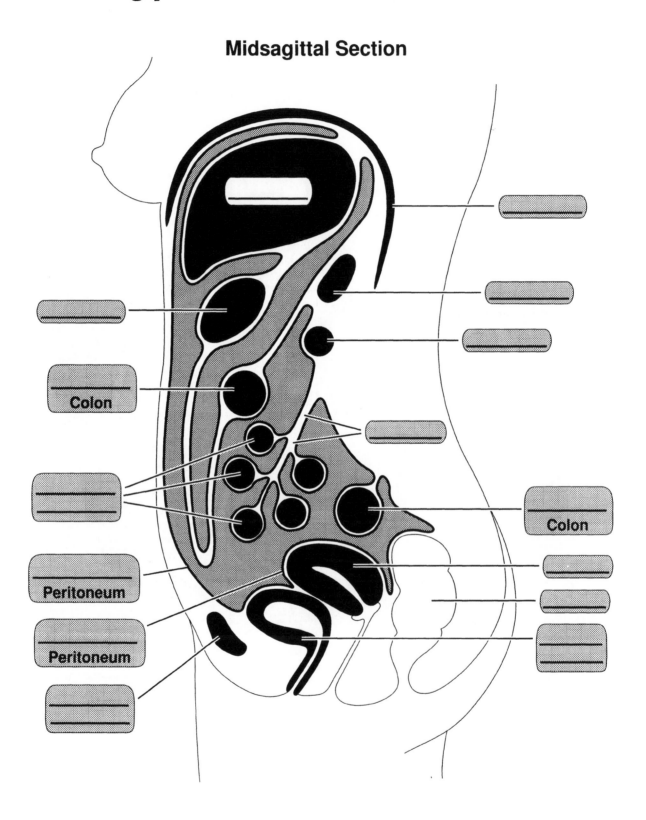

Colon

Colon

Peritoneum

Peritoneum

CHEMICAL FORMULAS
the most abundant elements in a cell

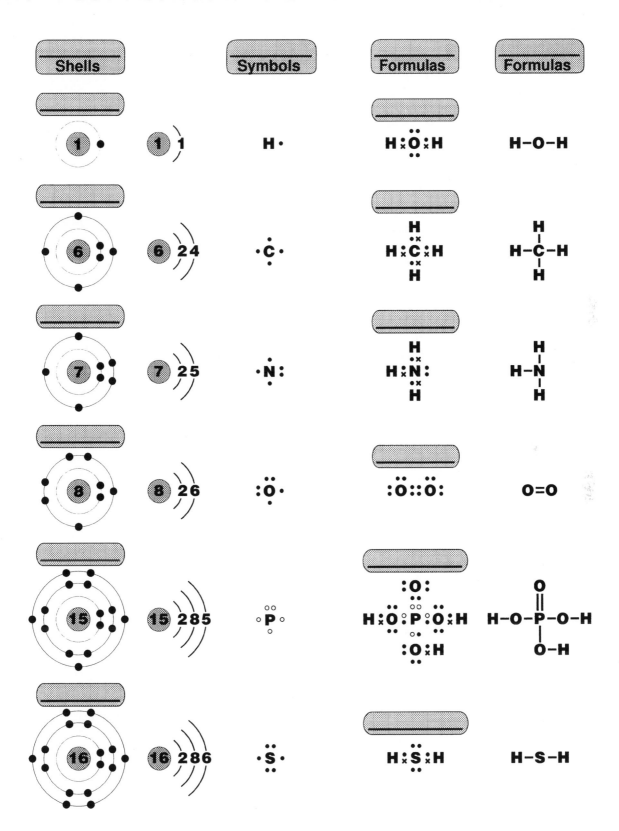

TYPES OF CHEMICAL BONDS

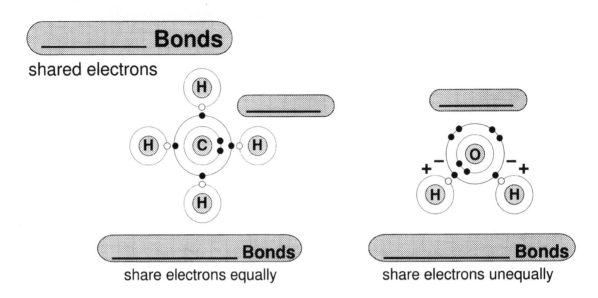

_____ **Bonds**

shared electrons

_____

_____ **Bonds**

share electrons equally

_____ **Bonds**

share electrons unequally

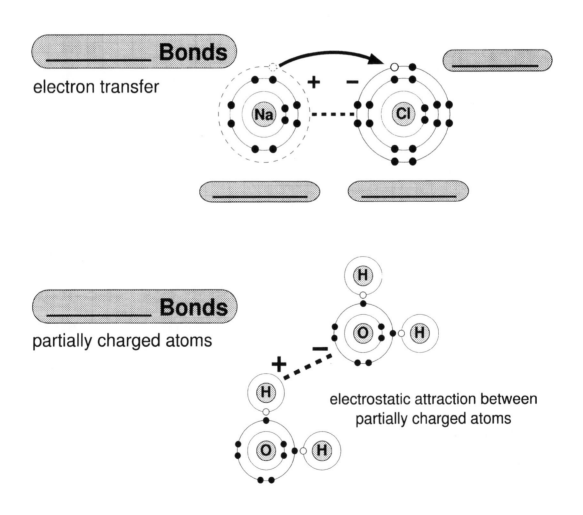

_____ **Bonds**

electron transfer

_____

_____

_____

_____ **Bonds**

partially charged atoms

electrostatic attraction between
partially charged atoms

TYPES OF CHEMICAL REACTIONS

_____ (_____ Reaction)

_____ (loss of a water molecule)

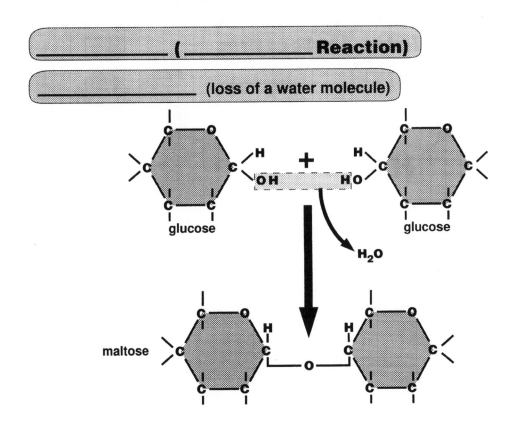

_____ (_____ Reaction)

_____ (addition of a water molecule)

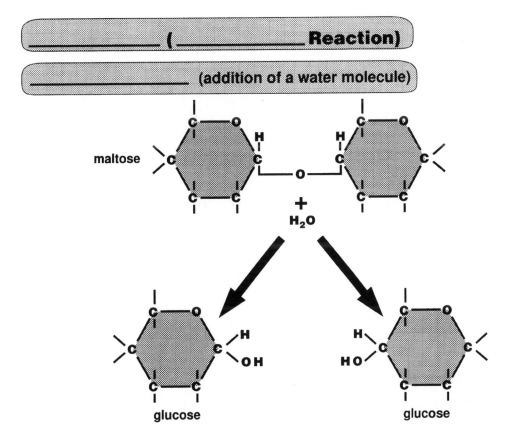

WATER AND ELECTROLYTES

Polarity of Water Molecules

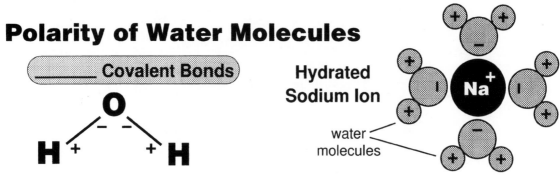

_____ Covalent Bonds

O
H$^+$ $^-$ $^-$ $^+$ H

Hydrated
Sodium Ion

Na$^+$

water
molecules

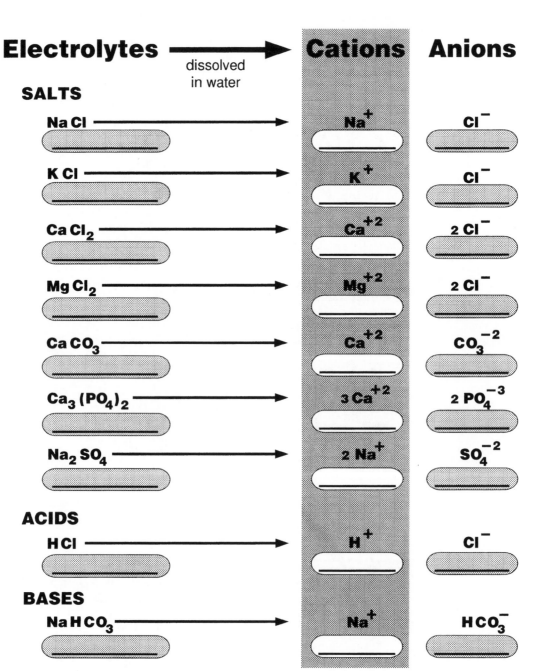

Electrolytes $\longrightarrow$ dissolved in water	Cations	Anions
SALTS		
Na Cl $\longrightarrow$	Na$^+$	Cl$^-$
K Cl $\longrightarrow$	K$^+$	Cl$^-$
Ca Cl$_2$ $\longrightarrow$	Ca^{+2}	2 Cl$^-$
Mg Cl$_2$ $\longrightarrow$	Mg^{+2}	2 Cl$^-$
Ca CO$_3$ $\longrightarrow$	Ca^{+2}	CO$_3^{-2}$
Ca$_3$ (PO$_4$)$_2$ $\longrightarrow$	3 Ca^{+2}	2 PO$_4^{-3}$
Na$_2$ SO$_4$ $\longrightarrow$	2 Na$^+$	SO$_4^{-2}$
ACIDS		
H Cl $\longrightarrow$	H$^+$	Cl$^-$
BASES		
Na H CO$_3$ $\longrightarrow$	Na$^+$	H CO$_3^-$

ORGANIC COMPOUNDS

Carbohydrates and lipids consist of carbon, hydrogen, and oxygen atoms.
Polysaccharides are polymers made up of chains of simple sugars such as glucose.

The main types of lipids are triglycerides, phospholipids, steroids, lipoproteins, and eicosanoids.

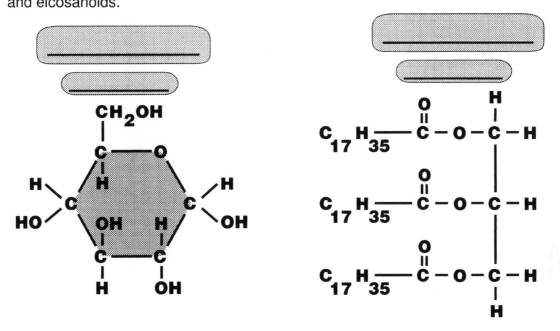

Proteins and nucleic acids contain nitrogen atoms (as well as carbon, hydrogen, & oxygen atoms), which distinguishes them from carbohydrates and lipids.

Proteins consist of chains of amino acids.
Nucleic acids consist of chains of nucleotides.

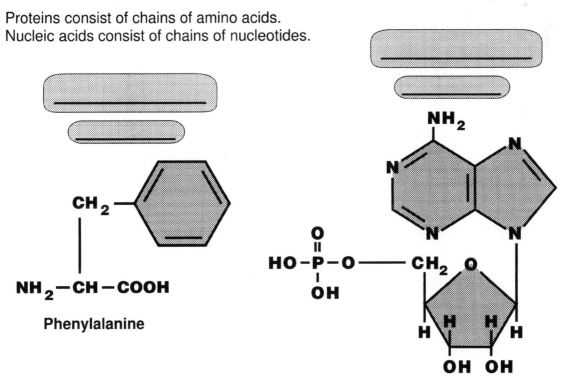

Phenylalanine

A Chain

B Chain

_____

(51 _____)

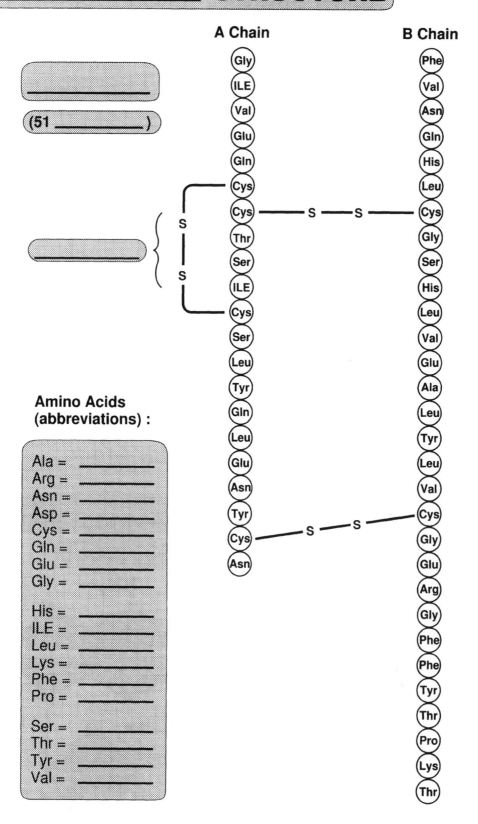

**Amino Acids
(abbreviations) :**

Ala = _____
Arg = _____
Asn = _____
Asp = _____
Cys = _____
Gln = _____
Glu = _____
Gly = _____

His = _____
ILE = _____
Leu = _____
Lys = _____
Phe = _____
Pro = _____

Ser = _____
Thr = _____
Tyr = _____
Val = _____

pH SCALE

As the acidity of a solution increases,
the pH ⬭_____⬭

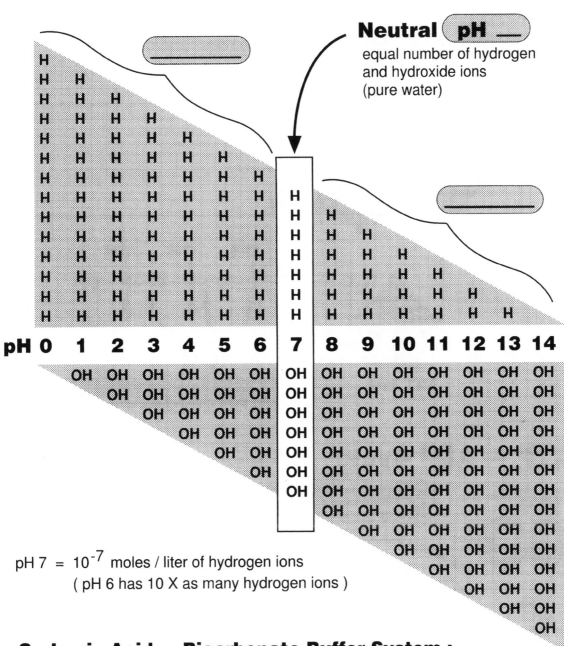

Neutral ⬭ **pH** __ ⬭
equal number of hydrogen
and hydroxide ions
(pure water)

pH 7 = 10^{-7} moles / liter of hydrogen ions
(pH 6 has 10 X as many hydrogen ions)

Carbonic Acid — Bicarbonate Buffer System :

Excess Hydrogen Ions (bicarbonate functions as a weak base) :

$$H^+ + HCO_3^- \text{ (weak base)} \rightleftharpoons H_2CO_3 \quad (\underline{\hspace{2cm}})$$

Shortage of Hydrogen Ions (carbonic acid functions as a weak acid) :

$$H_2CO_3 \text{ (weak acid)} \rightleftharpoons H^+ + HCO_3^- \quad (\underline{\hspace{2cm}})$$

87

ATP (adenosine triphosphate)

Energy-dependent activities of cells depend upon the energy stored in the high-energy phosphate bonds of ATP.

High-energy phosphate bonds are broken as energy is required by the cell.

Catabolism of fuels provides energy for the regeneration of ATP from ADP and inorganic phosphates.

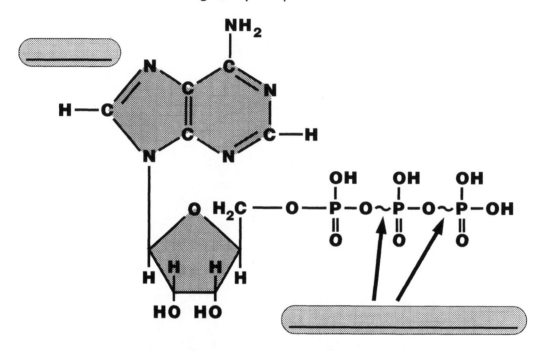

ATP — ADP Cycle

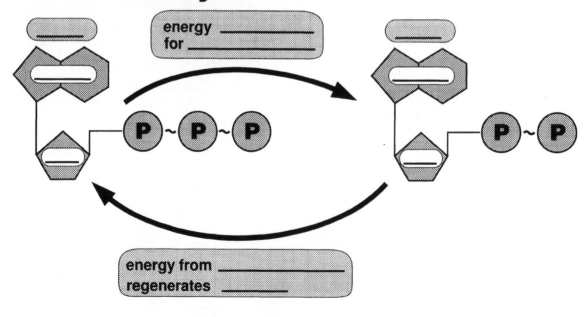

energy _____ for _____

energy from _____ regenerates _____

DIFFUSION AND OSMOSIS

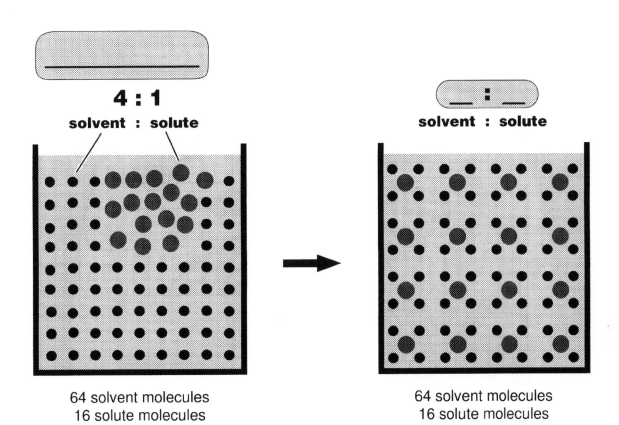

4 : 1
solvent : solute

solvent : solute

64 solvent molecules
16 solute molecules

64 solvent molecules
16 solute molecules

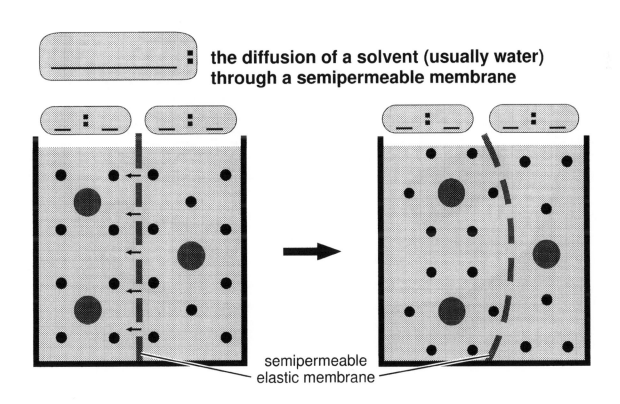

**the diffusion of a solvent (usually water)
through a semipermeable membrane**

semipermeable
elastic membrane

CELL STRUCTURES
Generalized Animal Cell

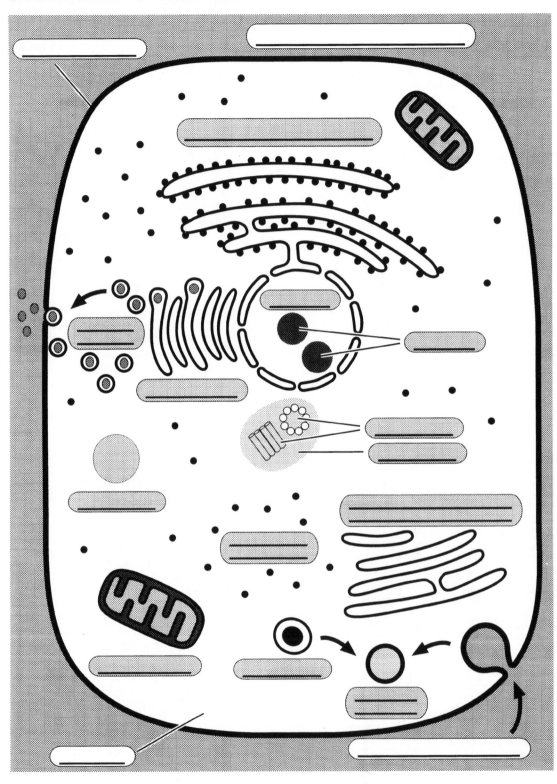

MEMBRANE TRANSPORT
Fluid Mosaic Model

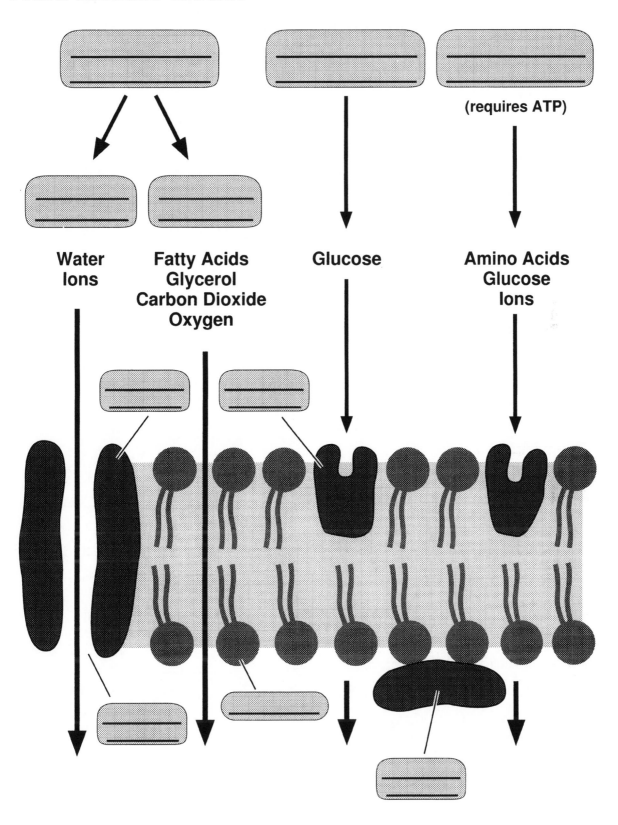

Water
Ions

Fatty Acids
Glycerol
Carbon Dioxide
Oxygen

Glucose

(requires ATP)

Amino Acids
Glucose
Ions

CELL DIVISION

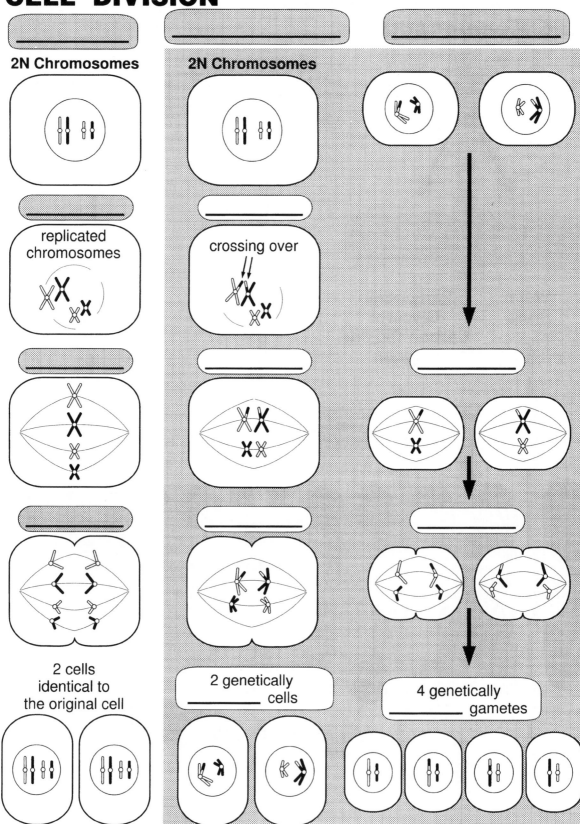

2N Chromosomes

2N Chromosomes

replicated
chromosomes

crossing over

2 cells
identical to
the original cell

2 genetically
_____ cells

4 genetically
_____ gametes

CELLULAR RESPIRATION
Overview

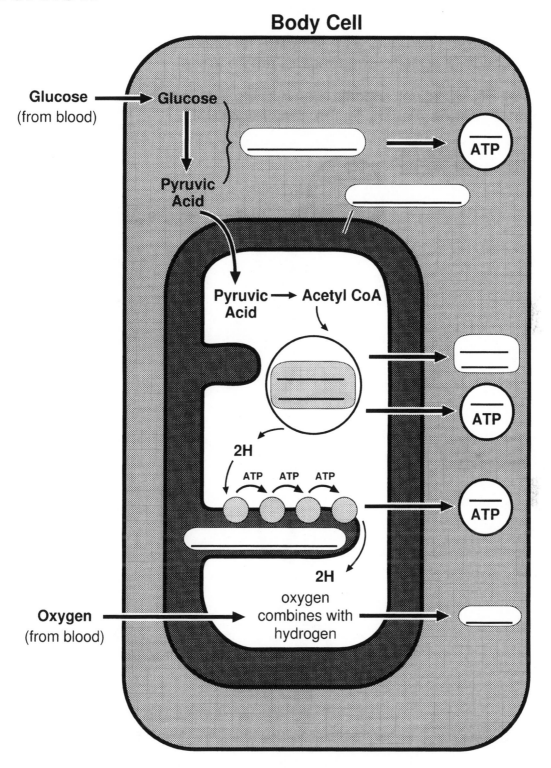

Body Cell

Glucose
(from blood)

Glucose

ATP

Pyruvic
Acid

Pyruvic → Acetyl CoA
Acid

ATP

2H

ATP   ATP   ATP

ATP

2H
oxygen
combines with
hydrogen

Oxygen
(from blood)

Glucose + Oxygen + 2 ATP ⟶ Carbon Dioxide + Water + 38 ATP

PROTEIN SYNTHESIS

A messenger RNA carries genetic information coded in its base sequences from the nucleus to a ribosome.

The messenger RNA acts as a template for building a polypeptide (a specific sequence of amino acids).

Transfer RNAs deliver specific amino acids that match the genetic codes present on the messenger RNA.

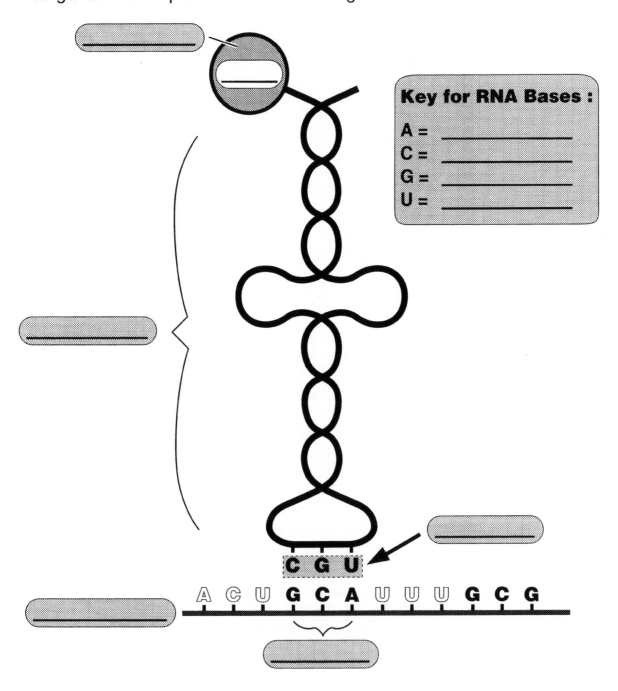

Key for RNA Bases :

A = _____
C = _____
G = _____
U = _____

C G U

A C U G C A U U U G C G

LIGANDS AND RECEPTORS
Example : hormone – receptor interaction

Receptors for peptide hormones are located on the outer surface of ⬭

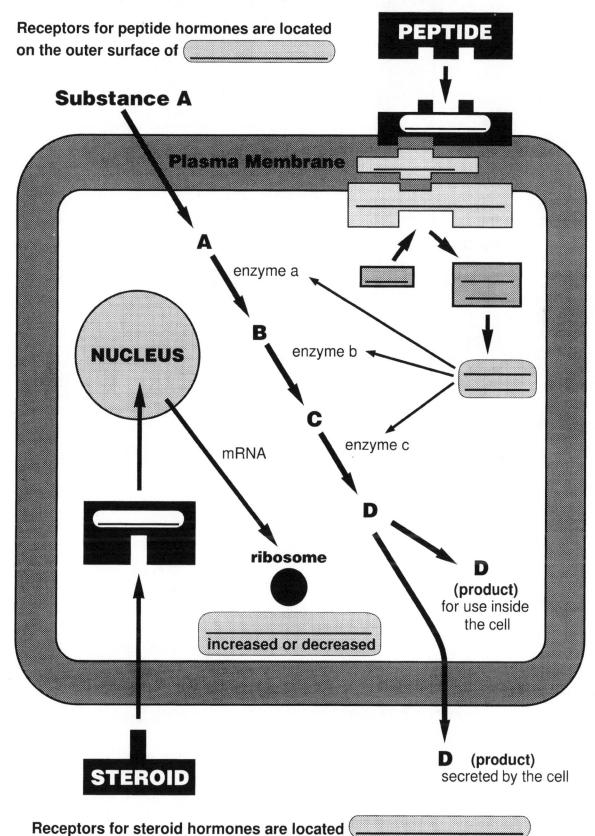

Receptors for steroid hormones are located ⬭

THE 4 BASIC TISSUES

_____ **Tissue**

Covering & Lining

Glandular

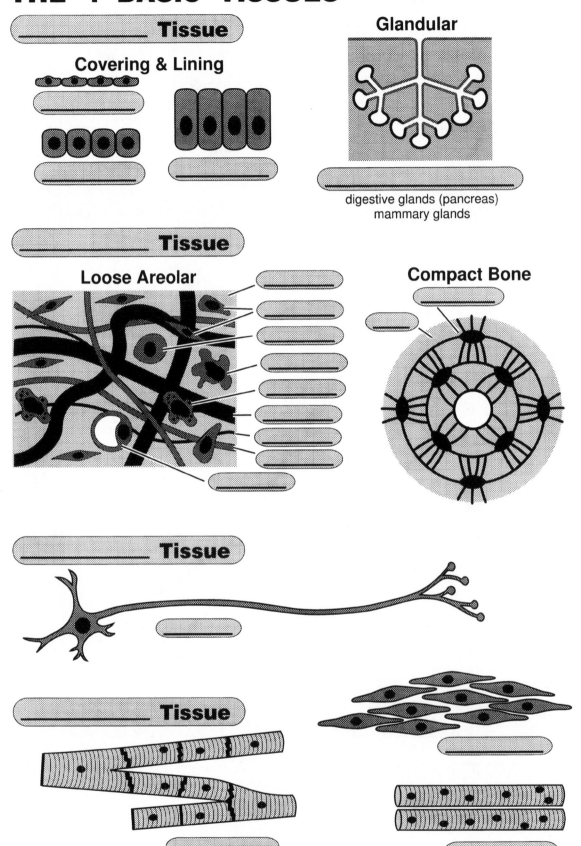

digestive glands (pancreas)
mammary glands

_____ **Tissue**

Loose Areolar

Compact Bone

_____ **Tissue**

_____ **Tissue**

EPITHELIAL TISSUES
Covering and Lining Epithelium

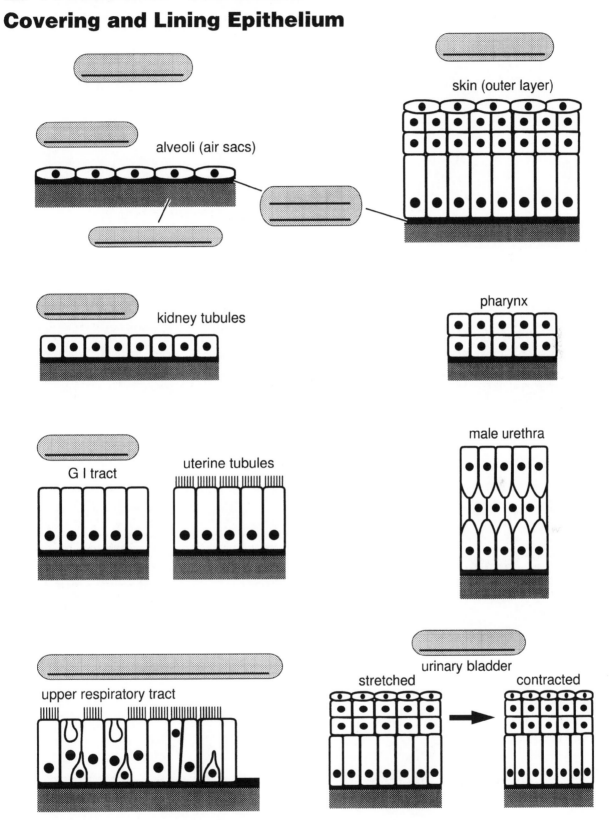

skin (outer layer)

alveoli (air sacs)

kidney tubules

pharynx

G I tract

uterine tubules

male urethra

upper respiratory tract

urinary bladder

stretched

contracted

CONNECTIVE TISSUES

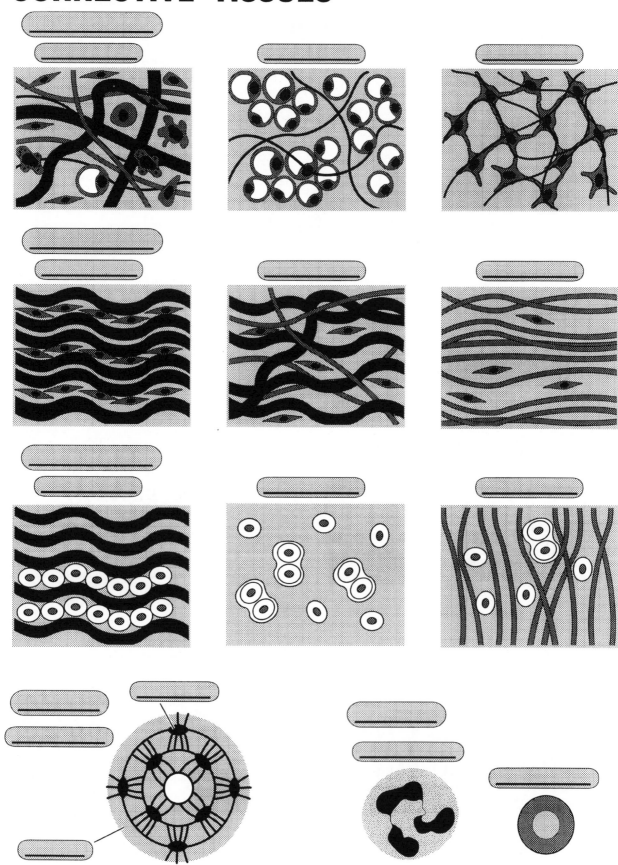

NERVOUS TISSUES
Neurons and Neuroglia

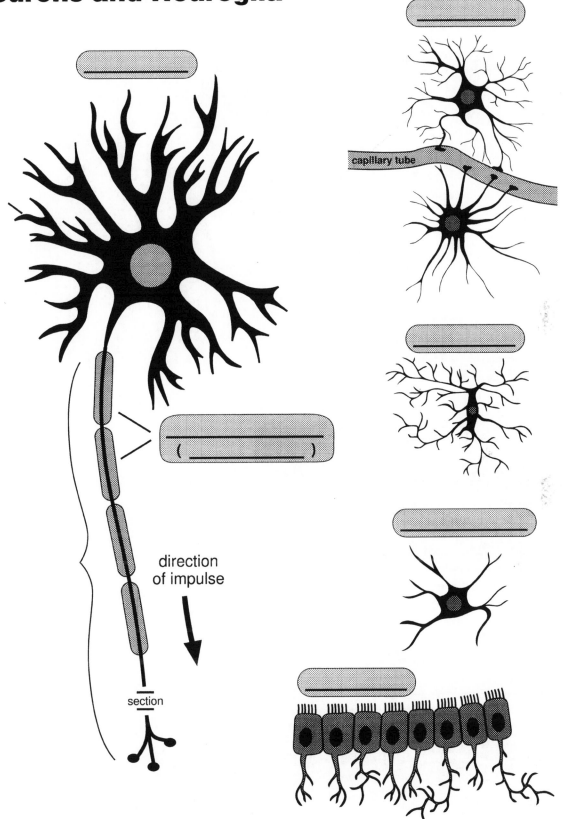

capillary tube

direction
of impulse

section

MUSCLE TISSUES

_____ Muscle

_____

SECTION

_____

_____
(cross section)

_____
(actin & myosin filaments inside)

_____

_____ Muscle

_____

|←——100 μm——→|

_____

_____

_____

_____ Muscle

_____

_____

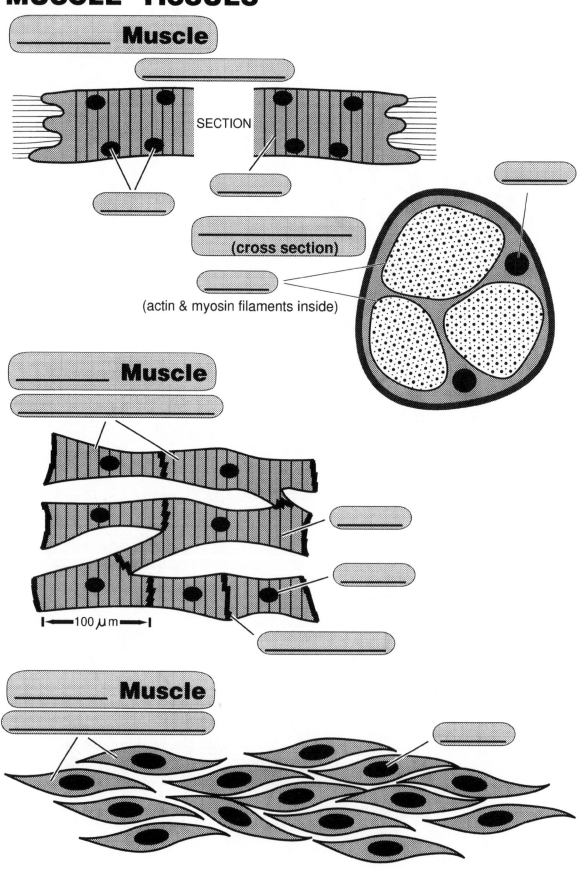

SKIN

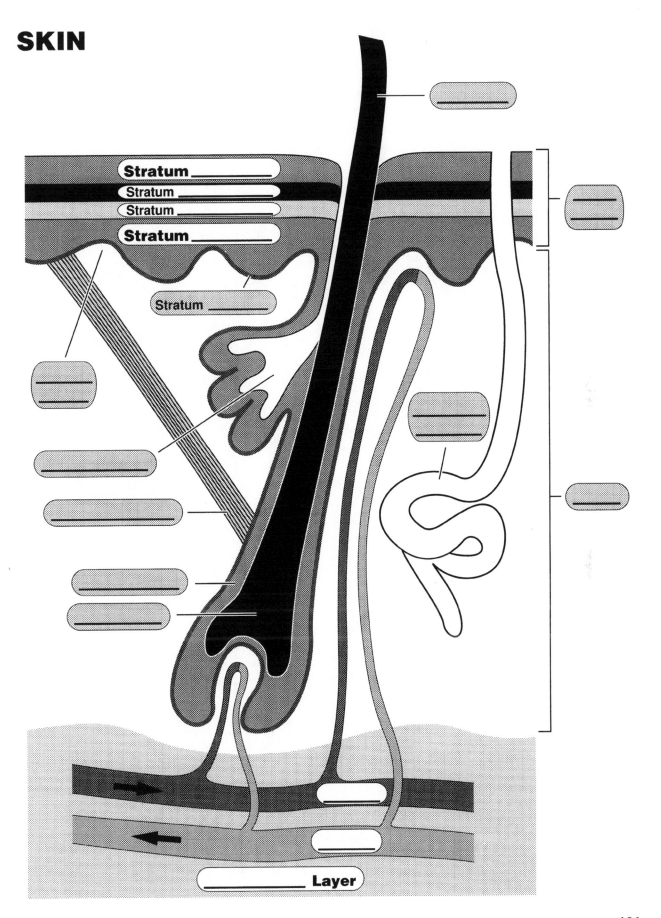

Stratum _____

Stratum _____

Stratum _____

Stratum _____

Stratum _____

_____ Layer

SKIN RECEPTORS
Touch, Temperature, Pain, and Pressure

Connective Tissue

Connective Tissue

_____ **Layer**

SURVIVAL OF INDIVIDUAL CELLS

Single-Celled Organisms
a water environment that is naturally (_____ & _____)

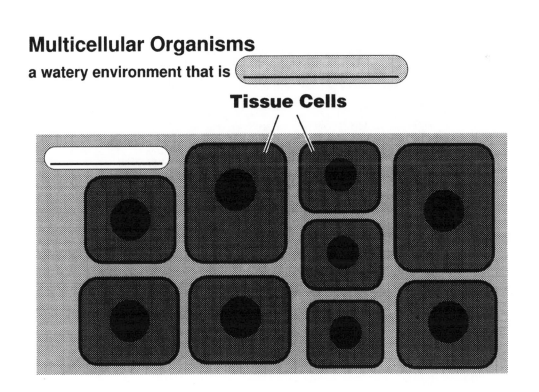

Water Environment

Amoeba

volume of water is (_____) relative to cell size

Multicellular Organisms
a watery environment that is (_____)

Tissue Cells

(_____)

FLUID COMPARTMENTS

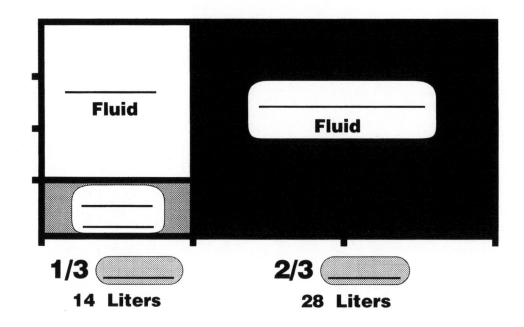

Fluid

Fluid

1/3 ⬭‾‾‾ **2/3** ⬭‾‾‾

14 Liters **28 Liters**

CAPILLARY BED

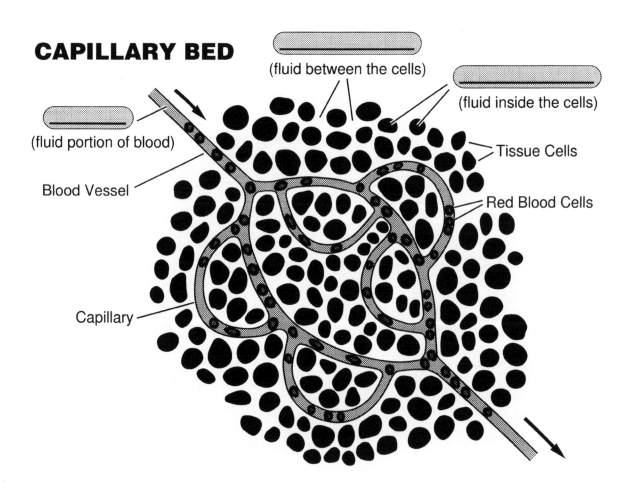

⬭‾‾‾
(fluid between the cells)

⬭‾‾‾
(fluid inside the cells)

⬭‾‾‾
(fluid portion of blood)

Tissue Cells

Blood Vessel

Red Blood Cells

Capillary

FEEDBACK SYSTEMS (Reflex Arcs)

A feedback system helps to maintain a stable internal environment.

The 3 basic components of a feedback system are : receptor, control center, and effector.

Thermostat

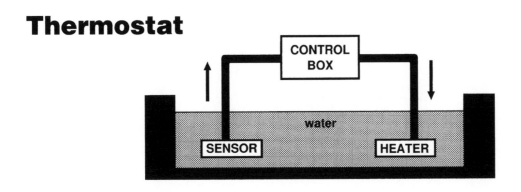

Feedback System
(nervous system responses)

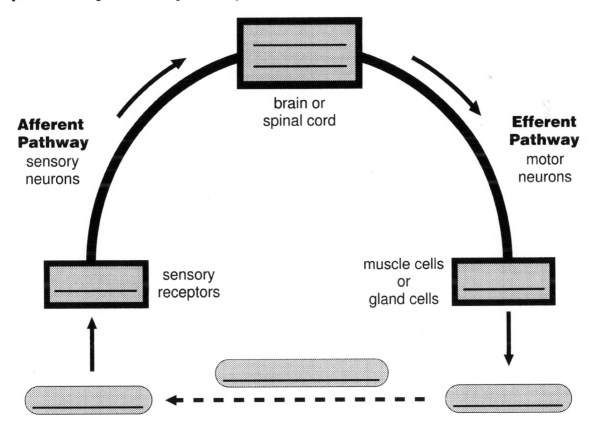

Afferent Pathway
sensory neurons

brain or spinal cord

Efferent Pathway
motor neurons

sensory receptors

muscle cells or gland cells

Part III : Terminology

Pronunciation Guide   108

Glossary   113

Bibliography   *123*

Pronunciation Guide

abdomen AB - dō - men	basal BĀ - sal
abdominal ab - DOM - i - nal	bicarbonate bī - KAR - bō - nāt
abdominopelvic ab - dom′ - i - nō - PEL - vik	brachial BRĀ - kē - al
acetyl AS - e - til	buccal BUK - al
acromial a - KRŌ - mē - al	bursa BER - sa
adenosine a - DEN - ō - sēn	bursae BER - sē
adenylate cyclase a - DEN - i - lāt SĪ - klās	
adipocyte AD - i - pō - sīt	calcaneal kal - KĀ - nē - al
adipose AD - i - pōs	canaliculus kan′ - a - LIK - yoo - lus
aerobic air - Ō - bik	carbonic kar - BON - ik
afferent AF - er - ent	carotene KAR - o - tēn
alkaline AL - ka - līn	carpal KAR - pal
allosteric al′ - ō - STER - ik	catabolic kat′ - a - BOL - ik
amine a - MĒN	catabolism ka - TAB - ō - lizm
amino a - MĒN - ō	catalyst KAT - a - list
anabolic an′ - a - BOL - ik	catalytic kat′ - a - LIT - ik
anabolism a - NAB - ō - lizm	cation KAT - ī - on
anaerobic an - air - Ō - bik	caudad KAW - dad
anaphase AN - a - fāz	centriole SEN - trē - ōl
anatomical an′ - a - TOM - i - kal	centromere SEN - trō - mēr
anion AN - ī - on	centrosome SEN - trō - sōm
antebrachial an′ - tē - BRĀ - kē - al	cephalic se - FAL - ik
antecubital an′ - tē - KYOO - bi - tal	cerebrospinal se - rē′ - brō - SPĪ - nal
anticodon an′ - ti - KŌ - don	ceruminous se - ROO - mi - nus
antiport AN - ti - port	cervical SER - vi - kul
apical AP - i - kal	cervix SER - viks
apoenzyme ap′ - ō - EN - zīm	cholesterol kō - LES - te - rol
arachidonic a - ra - ki - DON - ik	chondrocyte KON - drō - sīt
areolar a - RĒ - ō - lar	chromatid KRŌ - ma - tid
arrector pili a - REK - tor PI - lē	chromatin KRŌ - ma - tin
astrocyte AS - trō - sīt	chromosome KRŌ - mō - sōm
axilla ak - SIL - a	cilia SIL - ē - a
axon AK - son	codon KŌ - don

coenzyme kō - EN - zīm
cofactor KŌ - fak - tor
collagen KOL - a - jen
columnar kol - LUM - nar
conjunctiva kon - junk - TĪ - va
corneum KOR - nē - um
coronal kō - RŌ - nal
covalent kō - VĀL - ent
coxal KOK - sal
cranial KRĀ - nē - al
crista KRIS - ta
cristae KRIS - tē
crural KROOR - al
cuboidal kyoo - BOY - dal
cutaneous kyoo - TĀ - nē - us
cytokinesis sī′ - tō - ki - NĒ - sis
cytology sī - TOL - ō - jē
cytoplasm SĪ - tō - plazm
cytoskeleton sī - tō - SKEL - e - ton
cytosol SĪ - tō - sol

dehydration dē - hī - DRĀ - shun
dehydrogenase dē - HĪ - drō - je - nās
denaturation dē - nā - chur - Ā - shun
dendrite DEN - drīt
deoxyribonucleic dē - ok′ - sē - rī - bō - nyoo - KLĀ - ik
dermis DER - mis
diaphragm DĪ - a - fram
differentiation dif′ - e - ren′ - shē - Ā - shun
diffusion dif - YOO - zhun
digital DIJ - i - tal
diploid DIP - loyd
disaccharide dī - SAK - a - rīd
dissociation dis′ - sō - sē - Ā - shun
distal DIS - tal
dorsal DOR - sal
duodenum doo - ō - DĒ - num

ectoderm EK - tō - derm
effector e - FEK - tor
efferent EF - er - ent
eicosanoid ī - KŌ - sa - noid
electrolyte e - LEK - trō - līt
electronegativity e - lek′ - trō - neg′ - a - TIV - i - tē
embryology em′ - brē - OL - ō - jē
endocrine EN - dō - krin
endocytosis en′ - dō - sī - TŌ - sis
endoderm EN - dō - derm
endoplasmic en′ - dō - PLAS - mik
endothelial en′ - dō - THĒ - lē - al
endothelium en′ - dō - THĒ - lē - um
enzyme EN - zīm
ependymal e - PEN - de - mal
epidermal ep′ - i - DERM - al

epidermis ep′ - i - DERM - is
epithelial ep′ - i - THĒ - lē - al
epithelium ep′ - i - THĒ - lē - um
eponychium ep′ - ō - NIK - ē - um
erythrocyte e - RITH - rō - sīt
Eustachian yoo - STĀ - kē - an
exocrine EK - sō - krin
exocytosis ex′ - ō - sī - TŌ - sis
extracellular eks - tra - SEL - yoo - lar

facilitated fa - SIL - i - tā - ted
fascia FASH - ē - a
femoral FEM - or - al
fibroblast FĪ - brō - blast
fibrocartilage fī - brō - KAR - ti - lij
fructose FRUK - tōs

galactose ga - LAK - tōs
gamete GAM - ēt
ganglia GANG - glē - a
gene JĒN
glandular GLAN - dyoo - lar
gluteal GLOO - tē - al
glycerol GLIS - er - ol
glycolipid glī - kō - LIP - id
glycolysis glī - KOL - i - sis
Golgi GOL - jē
granulosum gran - yoo - LŌ - sum
groin groyn

hallux HAL - uks
haploid HAP - loyd
Haversian ha - VĒR - shun
histology hiss - TOL - ō - jē
homeostasis hō - mē - ō - STĀ - sis
homologous hō - MOL - ō - gus
hormone HOR - mōn
hyaline HĪ - a - līn
hydrochloric hī′ - drō - KLOR - ik
hydrolase HĪ - drō - lās
hydrolysis hī - DROL - i - sis
hypertonic hī′ - per - TON - ik
hyponychium hī′ - pō - NIK - ē - um
hypotonic hī′ - pō - TON - ik

inclusion in - KLOO - zhun
inguinal IN - gwi - nal
integral IN - te - gral
integumentary in - teg′ - yoo - MEN - tar - ē
intercalated in - TER - ka - lāt - ed
interphase IN - ter - fāz
interstitial in′ - ter - STISH - al
intracellular in′ - tra - SEL - yoo - lar
ion Ī - on

ionization   ī′-on-i-ZĀ-shun
isomerase   ī-SOM-er-ās
isotonic   ī′-sō-TON-ik
isotope   Ī-sō-tōp

keratin   KER-a-tin
keratinocyte   ker-a-TIN-ō-sīt
ketone   KĒ-tōn
kinase   KĪ-nās
kinetic   ki-NET-ik
Krebs   KREBZ

lactose   LAK-tōs
lacuna   la-KOO-na
lamellae   la-MEL-ē
lamina   LAM-i-na
leukocyte   LOO-kō-sīt
leukotriene   loo-kō-TRĪ-ēn
ligand   LĪ-gand
lipase   LIP-ās
lipid   LI-pid
lipoprotein   lip′-ō-PRŌ-tēn
lucidum   LOO-si-dum
lumbar   LUM-bar
lumen   LOO-men
lunula   LOO-nyoo-la
lymphatic   lim-FAT-ik
lymphoid   LIM-foyd
lysosome   LĪ-sō-sōm

macromolecule   mak′-rō-MOL-e-kyool
macrophage   MAK-rō-fāj
maltose   MAWL-tōs
mammary   MAM-ar-ē
matrix   MĀ-triks
medial   MĒ-dē-al
mediastinum   mē′-dē-as-TĪ-num
meiosis   mī-Ō-sis
Meissner   MĪS-ner
melanin   MEL-a-nin
melanocyte   MEL-a-nō-sīt
menisci   men-IS-ī
Merkel   MER-kel
merocrine   MER-ō-krin
mesenchyme   MEZ-en-kīm
mesentery   MEZ-en-ter′-ē
mesoderm   MEZ-ō-derm
mesothelium   mez′-ō-THĒ-lē-um
metabolism   me-TAB-ō-lizm
metabolite   me-TAB-ō-līt
metacarpal   met′-a-KAR-pal
metaphase   MET-a-fāz
metatarsal   met′-a-TAR-sal
microfilament   mī′-krō-FIL-a-ment

microglia   mī-KROG-lē-a
microtubule   mī′-krō-TOOB-yool
microvilli   mī′-krō-VIL-ē
midsagittal   mid-SAG-i-tal
mitochondria   mī′-tō-KON-drē-a
mitochondrion   mī′-tō-KON-drē-on
mitosis   mī-TŌ-sis
mitotic   mī-TOT-ik
mole   MŌL
molecule   MOL-e-kyool
monocyte   MON-ō-sīt
monomer   MON-ō-mer
monosaccharide   mon′-ō-SAK-a-rīd
mosaic   mō-ZĀ-ik
mucosa   myoo-KŌ-sa
mucous   MYOO-kus
mucus   MYOO-kus
myelin   MĪ-e-lin
myofibril   mī′-ō-FĪ-bril

nasal   NĀ-zal
neuroglia   noo-ROG-lē-a
neurolemmocyte   noo-rō-LEM-ō-sīt
neurotransmitter   noo-rō-TRANS-mit-e
neutron   NOO-tron
nitrogenous   nī-TROJ-e-nus
nuclei   NOO-klē-ī
nucleic   noo-KLĀ-ik
nucleoli   noo-KLĒ-ō-lī
nucleolus   noo-KLĒ-ō-lus
nucleotide   NOO-klē-ō-tīd

oblique   ō-BLĒK
olecranal   ō-LEK-ra-nal
oligodendrocyte   OL-i-gō-den′-drō-sīt
oral   Ō-ral
osmosis   os-MŌ-sis
osseous   OS-ē-us
osteocyte   OS-tē-ō-sīt′
osteon   OS-tē-on
otic   Ō-tik
oxidase   OK-si-dās

palmar   PAL-mar
papilla   pa-PIL-a
papillae   pa-PIL-ē
papillary   PAP-i-lar′-ē
paracrine   PAR-a-krin
paranasal   par-a-NĀ-zal
parasagittal   par-a-SAG-i-tal
parenchyma   pa-REN-ki-ma
parietal   pa-RĪ-e-tal
patellar   pa-TELL-ar
pectoral   PEK-tō-ral

pedal PED-al
pelvic PEL-vik
peptide PEP-tīd
pericardial per'-i-KAR-dē-al
perikaryon per'-i-KAR-ē-on
peritoneum per'-i-tō-NĒ-um
peroxisome pe-ROKS-i-sōm
phagocyte FAG-ō-sīt
phagocytic fag'-ō-SIT-ik
phagocytosis fag'-ō-sī-TŌ-sis
phalangeal fah-LAN-jē-al
phosphate FOS-fāt
phospholipid fos'-fō-LIP-id
phosphorylation fos'-for-i-LĀ-shun
physiology fiz'-ē-OL-ō-jē
pilus PĪ-lus
pinocytosis pi'-nō-sī-TŌ-sis
plantar PLAN-tar
plasma PLAZ-ma
pleura PLOOR-a
pleural PLOOR-al
plexus PLEK-sus
pollex POL-eks
polyhedral pol'-ē-HĒ-dral
polymer POL-i-mer
polypeptide pol'-ē-PEP-tīd
polysaccharide pol'-ē-SAK-a-rīd
popliteal pop-LIT-ē-al
proliferation prō-lif'-er-Ā-shun
prophase PRŌ-fāz
propria PRŌ-prē-a
prostaglandin pros'-ta-GLAN-din
prostate PROS-tāt
protease PRŌ-tē-ās
protein PRŌ-tēn
proton PRŌ-ton
proximal PROK-si-mal
pseudopod SOO-dō-pod
pseudostratified soo'-dō-STRAT-i-fīd
pubic PYOO-bik
pyruvic pī-ROO-vik

quadrant KWAD-rant
quaternary KWA-ter-ner'-ē

Ranvier ron-VĒ-ā
reactant rē-AK-tant
renal RĒ-nal
replicate REP-li-kāt
reticular re-TIK-yoo-lar
reticulum re-TIK-yoo-lum
ribonucleic rī-bō-noo-KLĀ-ik
ribosomal rī-bō-SŌM-al
ribosome RĪ-bō-sōm

sagittal SAJ-i-tal
salivary SAL-i-ver-ē
Schwann SCHVON
sebaceous se-BĀ-shus
serosa sir-Ō-sa
serous SIR-us
sigmoid SIG-moyd
sinus SĪ-nus
solute SOL-yoot
solvent SOL-vent
somatic sō-MAT-ik
spinosum spin-Ō-sum
squamous SKWĀ-mus
steroid STER-oyd
stratum STRĀ-tum
stroma STRŌ-ma
subcutaneous sub'-kyoo-TĀ-nē-us
subserous sub-SER-us
substrate SUB-strāt
sucrose SOO-krōs
sudoriferous soo'-dor-IF-er-us
sural SOO-ral
symphysis SIM-fi-sis
symport SIM-port
synapsis sin-AP-sis
synovial si-NŌ-vē-al

tarsal TAR-sal
telophase TEL-ō-fāz
tertiary TUR-shē-er'-ē
thoracic thō-RAS-ik
trabecula tra-BEK-yoo-la
trabeculae tra-BEK-yoo-lē
transferase TRANS-fer-ās
triglyceride trī-GLIS-er-īd
triphosphate trī-FOS-fāt

umbilical um-BIL-i-kul
ureter YOO-re-ter
urethra yoo-RĒ-thra
urinary YOO-ri-ner-ē
uterus YOO-ter-us

valence VĀ-lens
vaporization vā-por-i-ZĀ-shun
vascular VAS-kyoo-lar
ventral VEN-tral
viscera VIS-er-a
visceral VIS-er-al
volar VŌ-lar

zygote ZĪ-gōt
zymogenic zī-mō-JEN-ik

Glossary of Terms

Abdominal cavity   Superior portion of the abdominopelvic cavity. It contains the stomach, spleen, liver, gallbladder, pancreas, small intestine, and most of the large intestine.

Abdominopelvic cavity   Inferior component of the ventral body cavity. It is subdivided into an upper abdominal cavity and a lower pelvic cavity.

Abdominopelvic quadrants   Four regions of the abdominopelvic cavity. It is formed by two perpedicular imaginary lines through the umbilicus. Includes the right and left upper quadrants and the right and left lower quadrants.

Abdominopelvic regions   Nine regions of the abdominopelvic cavity formed by four imaginary lines. They include : the right & left hypochondriac; right & left lumbar; right & left iliac; epigastric; umbilical; and pubic regions.

Absorption   The intake of fluids or other substances by cells of the skin or mucous membranes; the passage of digested foods from the gastrointestinal tract into blood or lymph.

Acetyl coenzyme A   A molecule inside a mitochondrion that transfers acetyl groups to the Krebs cycle.

Acid   A proton donor, or substance that dissociates into hydrogen ions (H^+) and anions. It is characterized by an excess of hydrogen ions and a pH less than 7.

Acromial   Pertaining to the acromion.

Acromion   Portion of the scapula (shoulder blade) that forms the most prominent point of the shoulder.

Activation energy   The minimum amount of energy required for a chemical reaction to occur. A match supplies the activation energy needed to start a fire.

Active site   The region of an enzyme that interacts with a substrate molecule. Some active sites are rigid, while others change shape to fit snugly around the substrate.

Active transport   The movement of substances across cell membranes against a concentration gradient; requires the expenditure of energy (ATP).

Adenosine diphosphate (ADP)   The product formed when ATP is broken down to release energy; contains two phosphate groups.

Adenosine triphosphate (ATP)   The energy "currency" of all living cells; energy readily available for cellular activities. It is synthesized in a process called cellular respiration.

Adenylate cyclase   An enzyme that catalyzes the transformation of ATP to cyclic AMP. It is located on the inner surface of plasma membranes.

Adipocyte   Fat cell; stores triglycerides as an energy reserve for ATP production. It is derived from a fibroblast.

Aerobic respiration   The sequence of reactions in cellular respiration that cannot occur unless oxygen is present; includes Krebs cycle and the electron transport system.

Alkaline   A solution that has a hydrogen ion concentration lower than that of pure water; a pH greater than 7. Also called *basic*.

Allosteric modulation   When the binding of a ligand to a protein receptor alters the shape of a second active site on the receptor.

Amino acids   The building blocks (monomers) of proteins; chains of amino acids form proteins. Each amino acid consists of a central carbon atom, an acidic carboxyl group (—COOH) , a basic amino group (—NH_2), and a variable R group. There are 20 different amino acids.

Amino group   The portion of an amino acid consisting of a nitrogen atom and two hydrogen atoms (—NH_2); it can act as a weak base, accepting a hydrogen ion (proton).

Anabolism   Includes all synthetic energy-requiring reactions that take place in cells.

Anaerobic respiration   Glycolysis. The sequence of reactions in cellular respiration that do not require the presence of oxygen. During this process one glucose molecule is converted into two pyruvic acid molecules.

Anaphase   The third stage of mitosis; chromatids that have separated at the centromeres move to opposite poles.

Anatomical position   A position of the body universally used in anatomical descriptions. The body is erect, facing the observer, the upper extremities are at the sides, the palms of the hands are facing forward, and the feet are on the floor.

Anatomy   The structures of the body; the study of body structures.

Anion   A negatively charged ion. An example is the chloride ion (Cl^-).

Antebrachial   Of or relating to the forearm.

Antecubital   Located in front of the elbow.

Anterior   Nearer to the front; opposite of posterior. Also called *ventral*.

Anticodon   Sequence of three bases on a transfer RNA. They attach to the matching codon of a messenger RNA during protein synthesis.

Antiport   Process by which two substances, usually Na^+ and another substance, move in opposite directions across a plasma membrane. Also called *countertransport*.

Apoenzyme   The protein portion of an enzyme molecule. (The nonprotein portion is called the cofactor.)

Arrector pili   Smooth muscles attached to hair follicles; contraction pulls the hairs into a more vertical position, resulting in "goose bumps."

Atom   The smallest particle of an element that still has the properties of the element. For example, if an atom of carbon is split, the resulting particles are no longer carbon. Chemical reactions involve the forming and breaking of bonds between atoms. An atom consists of a complex arrangement of negatively charged electrons orbiting around a positively charged nucleus; it is electrically neutral.

Atomic nucleus   The dense center of an atom, consisting of two types of subatomic particles : positively charged protons and neutral neutrons.

Atomic number   Number of protons or electrons in an atom.

Atomic weight   Total number of protons and neutrons in an atom.

Autosome   Any chromosome other than the pair of sex chromosomes. In humans, 44 of the 46 chromosomes are autosomes.

Axilla   The small hollow beneath the arm where it joins the body at the shoulders. Also called the *armpit*.

Axillary   Pertaining to the armpit.

Axon   The usually single, long process of a nerve cell that carries a nerve impulse away from the cell body.

Base   A proton acceptor; characterized by an excess of hydroxide ions (OH⁻) and a pH greater than 7. A base can also be a nitrogen-containing organic molecule; one of the three components of a nucleotide.

Basement membrane   A layer of extracellular material that attaches epithelial tissue to the underlying connective tissue; consists of the basal lamina and reticular lamina.

Blood tissue   A type of connective tissue. Consists of formed elements (red blood cells, white blood cells, platelets) and plasma. Also called *vascular tissue*.

Body cavity   A confined space within the body.

Bone tissue   A type of connective tissue. The two main kinds of bone tissue are compact bone and spongy bone. Also called *osseous tissue*.

Brachial   Relating to the arm; the region between the shoulder and the elbow (the "upper arm").

Buffer system   Chemicals that keep the pH of body fluids relatively constant. An example is the carbonic acid— bicarbonate buffer system : when there is a shortage of hydrogen ions, carbonic acid releases them; when there is an excess of hydrogen ions, bicarbonate combines with them.

Bulk flow   The movement of a fluid (liquid or gas) that results from a hydrostatic pressure gradient. Examples are the flow of blood and the flow of air in and out of the lungs.

Bulk transport   The movement of materials into a cell by endocytosis or out of a cell by exocytosis. Both processes require energy from ATP.

Calcaneal   Pertaining to the calcaneus (heel bone).

Callus   Abnormal thickening of the epidermis caused by prolonged exposure to friction or pressure.

Carbohydrate   An organic compound containing carbon, hydrogen, and oxygen; usually in a ratio of 1 : 2 : 1. Includes monosaccharides, disaccharides, and polysaccharides.

Carbonic acid—bicarbonate buffer system   An important regulator of blood pH. The bicarbonate ion can act as a weak base; carbonic acid can act as a weak acid.

Carboxyl group   The portion of an amino acid consisting of a carbon atom bound to an oxygen atom by a double bond and to a hydroxyl group by a single bond (—COOH); it can act as a weak acid, releasing a hydrogen ion (proton).

Cardiac muscle tissue   Heart muscle cells; characterized by branching, striated fibers attached by intercalated discs (thickening of the plasma membranes).

Carotene   A type of lipid present in egg yolk, carrots, and tomatoes. Vitamin A is synthesized from carotenes.

Carpal   Relating to the carpus (bones of the wrist).

Cartilage   A type of connective tissue. Consists of cells called chondrocytes, which are located in spaces called lacunae; the matrix consists of a dense network of collagenous and elastic fibers and a ground substance of chondroitin sulfate.

Catabolism   Chemical reactions that break down complex organic compounds into simpler ones with the release of energy.

Catalyst   A chemical that alters the rate of a chemical reaction without being changed in the process; usually present in small amounts. Catalysts for reactions that take place in cells are called enzymes.

Cation   A positively charged ion. An example is a sodium ion (Na⁺).

Caudal   *See* Inferior.

Cell   The basic structural and functional unit of all living things; the smallest structure capable of performing all the activities vital to life.

Cell division   Process by which a cell reproduces itself. It consists of a nuclear division (mitosis or meiosis) and a cytoplasmic division (cytokinesis). There are two main types : somatic and reproductive cell division.

Cellular respiration   The chemical reactions involved in the production of ATP by cells. It occurs in two major stages: (1) anaerobic respiration (glycolysis) and (2) aerobic respiration (Krebs cycle and the electron transport system).

Central canal   Found in compact bone tissue. A circular channel running longitudinally in the center of an osteon (Haversian system), containing blood and lymphatic vessels and nerves. Also called an *Haversian canal*.

Centrioles   Paired, cylindrical structures each consisting of a ring of microtubules and arranged at right angles to each other. Found within a centrosome. Participate in cell division.

Centromere   The clear, constricted portion of a chromosome where the two chromatids are joined; serves as the point of attachment for the chromosomal microtubules.

Centrosome   A dense area of cytoplasm near the nucleus of a cell; contains a pair of centrioles. During cell division it forms the mitotic spindle.

Cephalic   Relating to the head.

Ceruminous gland   A modified sweat gland in the external auditory canal whose secretions and those of a sebaceous (oil) gland combine to form cerumen (ear wax).

Cervical   Pertaining to the neck.

Chemical bond   Force of attraction that holds two atoms together in a molecule. Types include ionic, covalent, and hydrogen bonds.

Chemical element   Unit of matter that cannot be decomposed into a simpler substance by ordinary chemical reactions. Examples include hydrogen (H), carbon (C), oxygen (O).

Chemical energy   The energy stored in chemical bonds. When bonds are broken, energy is released; the formation of new bonds requires energy. Organic compounds that store large amounts of energy are called fuels; they contain many hydrogen-carbon bonds.

Chemical formula   A shorthand for describing the composition and structure of a molecule. *Molecular formulas* use chemical symbols of elements and numerical subscripts to give the proportions of the elements present in a compound. *Structural formulas* use a dash between chemical symbols to show the location of covalent bonds.

Chemical reaction   An interaction between substances in which chemical bonds are broken or formed (or both). There are four basic types of chemical reactions : synthesis, decomposition, exchange, and reversible reactions.

Chemical symbol   A shorthand abbreviation for an element. It consists of the first letter (or first two letters) in the English or Latin name of the element.

Cholesterol   The most abundant steroid (type of lipid) in animal tissues; located in cell membranes and used for the synthesis of steroid hormones and bile salts.

Chondrocyte   Cell of mature cartilage.

Chromatid   One of a pair of identical chromosomes attached by a centromere. They result from chromosome replication that occurs between cell divisions (during interphase).

Chromatin   The threadlike mass of the genetic material consisting principally of DNA. It is present in the nucleus of a nondividing cell.

Chromosome One of the 46 small, dark-staining bodies that appear in the nucleus of a human diploid (2n) cell during cell division; contain the genetic material (genes).

Cilium (plural : cilia) A hair or hairlike process projecting from a cell; its function is to move substances along the surface of the cell.

Codon Sequence of three bases on a messenger RNA. The codons determine the sequence of amino acids in the polypeptide being synthesized; each type of codon calls for a specific anticodon on a transfer RNA that carries a specific amino acid.

Coenzyme A nonprotein organic molecule that is associated with and activates an enzyme; many are derived from vitamins. Coenzymes and metal ions are called cofactors.

Cofactor The nonprotein portion of an enzyme. There are two types: coenzymes and metal ions.

Collagen A type of protein fiber found in most types of connective tissues. Abundant in bone, cartilage, tendons, and ligaments; gives strength to connective tissue.

Compound A substance that can be broken down into two or more other substances by chemical means.

Connective tissue The most abundant of the four basic tissue types in the body. It binds together, supports, and strengthens other tissues; protects and insulates internal organs; and compartmentalizes structures such as skeletal muscles.

Control center One of the three basic components of a feedback system (or reflex arc). Control centers receive information about changes in the environment from receptors and send out instructions to effectors (muscles or glands) that respond appropriately.

Coronal plane *See* Frontal plane.

Corpuscle of touch A sensory receptor for the sensation of touch; found in the dermal papillae of the skin, especially in palms and soles. Also called a *Meissner's corpuscle*.

Covalent bond A chemical bond in which valence electrons are shared.

Covalent modulation Protein receptors are altered (modulated) when a phosphate group attaches by covalent bonds.

Cranial cavity The subdivision of the dorsal body cavity formed by the cranial bones; contains the brain.

Crista (plural : cristae) A fold of the inner membrane of a mitochondrion. Cristae extend into the matrix and contain the enzymes needed for the electron transport system (ATP production). A crista is also a structure in the internal ear that serves as a receptor for dynamic equilibrium.

Crossing-over The exchange of a portion of one chromatid with another in a tetrad during meiosis. It permits an exchange of genes among otherwise identical chromatids. It is one factor that causes genetic variation.

Crural Pertaining to the leg; the region between the knee and the foot.

Cutaneous Pertaining to the skin.

Cyclic AMP (cAMP) Adenosine — 3', 5' — monophosphate. Molecule formed from ATP by the action of the enzyme adenylate cyclase; serves as an intracellular messenger (second messenger) for some hormones.

Cytokinesis Division of the cytoplasm during cell division.

Cytology The study of cells.

Cytoplasm All cellular contents located between the plasma membrane and the nucleus.

Cytoskeleton Filaments associated with cellular movement and shape. The three main types are microfilaments,

microtubules, and intermediate filaments.

Cytosol Fluid portion of the cytoplasm in which cell organelles are suspended.

Decomposition reaction A reaction in which a molecule breaks into smaller parts. Decomposition reactions in cells are called catabolic reactions (catabolism).

Deep Away from the surface of the body.

Deep fascia A sheet of connective tissue wrapped around a muscle to hold it in place.

Dehydration synthesis Formation of a chemical bond with the release of a water molecule.

Denaturation Change in the shape of a protein by heat, changes in pH, or other physical or chemical methods; the protein loses its physical and biological properties.

Dendrite A nerve cell process that carries a nerve impulse toward the neuron cell body.

Dense connective tissue Contains more numerous and thicker fibers but considerably fewer cells than loose connective tissue. Types include regular, irregular, and elastic.

Deoxyribonucleic acid (DNA) A double-stranded nucleic acid in the shape of a double helix, consisting of repeating units called nucleotides. The nucleotides are composed of the sugar deoxyribose, a phosphate group, and one of four nitrogenous bases (adenine, cytosine, guanine, or thymine). Genetic information is encoded in the sequence of the nucleotides.

Dermal papilla Fingerlike projection of the papillary region of the dermis of the skin; may contain blood capillaries or corpuscles of touch (Meissner's corpuscles).

Dermis A layer of dense connective tissue lying deep to the epidermis; the true skin or corium.

Diffusion A passive process in which there is a net movement of molecules or ions from a region of high concentration to a region of low concentration until equilibrium is reached.

Digital Pertaining to fingers or toes. Also called *phalangeal*.

Diploid The number of chromosomes characteristically found in the somatic cells of an organism. Symbolized 2n.

Disaccharide Double sugars. Maltose (two glucoses); sucrose (glucose & fructose); lactose (glucose & galactose).

Dissociation Separation of inorganic acids, bases, and salts into ions when dissolved in water. Also called *ionization*.

Distal Farther from the point of origin; toward the hand of the upper extremity or toward the foot of the lower extremity.

Dorsal Nearer to the back; opposite of ventral. Also called *posterior*.

Dorsal body cavity Cavity near the back (dorsal) surface of the body; consists of a cranial cavity and vertebral canal.

Ectoderm The outermost of the three primary germ layers; gives rise to the nervous system, the epidermis of the skin, and epidermal derivatives (hair, glands, and nails).

Effector One of the three basic components of a feedback system (reflex arc). A muscle or a gland, that responds to instructions received from a control center.

Eicosanoid A class of lipid derived from arachidonic acid; includes prostaglandins and leukotrienes.

Electrolyte Any compound that separates into ions when dissolved in water.

Electron A subatomic particle that carries a negative charge

115

and revolves around the nucleus of an atom.

Electronegativity   The degree of attraction that an atomic nucleus has for its electrons.

Electron transport system   A series of reactions that take energy from energy-rich electrons and use it to form high-energy phosphate bonds in ATP. The reactions occur on the cristae of mitochondria and require oxygen.

Endocrine gland   A gland that secretes hormones into the blood; a ductless gland.

Endocytosis   The uptake of large molecules and particles into a cell. A segment of plasma membrane surrounds the substance, encloses it, and brings it into the cell. Endocytosis includes phagocytosis, pinocytosis, and receptor-mediated endocytosis. All require the energy of ATP.

Endoderm   The innermost of the three primary germ layers of the developing embryo; it gives rise to the gastrointestinal and respiratory tracts, urinary bladder, and urethra.

Endoplasmic reticulum (ER)   A network of branched tubules and flattened sacs running through the cytoplasm of a cell. Its functions include synthesis and packaging of molecules, intracellular transport of molecules, storage of molecules, and support. Portions of ER where ribosomes are attached to the outer surface are called rough (granular) reticulum; portions that have no ribosomes are called smooth (agranular) reticulum.

Endothelium   The layer of simple squamous epithelium that lines the cavities of the heart, blood vessels, and lymphatic vessels.

Energy   The capacity to do work.

Enzyme   A substance that affects the speed of chemical reactions in cells; an organic catalyst.

Ependymal cells   Cells that line the fluid-filled cavities (ventricles) of the brain and probably assist in the circulation of cerebrospinal fluid.

Epidermal derivatives   Structures that develop from the embryonic epidermis. They include hair, nails, and skin glands (sebaceous, sudoriferous, and ceruminous).

Epidermis   The outer layer of skin composed of stratified squamous epithelium. The 5 layers (strata) from deepest to most superficial are: basale, spinosum, granulosum, lucidum, and corneum. The 4 principal cell types are: keratinocytes, melanocytes, Langerhans cells, and Merkel cells.

Epithelial tissue   Divided into two types. (1) covering and lining epithelium : forms the outer layer of the skin and the outer layer of some internal organs; forms the inner lining of blood vessels, ducts, body cavities; forms the interiors of respiratory, digestive, urinary, and reproductive systems. (2) glandular epithelium : includes the secreting portions of exocrine glands and endocrine glands.

Eponychium   Narrow band of stratum corneum at the proximal border of a nail that extends from the margin of the nail wall. Also called the *cuticle*.

Erythrocyte   Red blood cell.

Exchange reactions   A type of chemical reaction in which bonds are broken and new bonds are formed; one atom or group of atoms is replaced by another atom or group of atoms.

Exocrine gland   A cluster of epithelial cells that secretes substances (such as oil, sweat, mucus, or digestive enzymes) into ducts that lead to an epithelial surface.

Exocytosis   A process of discharging cellular products too large to go through the membrane. Particles for export are enclosed by Golgi membranes when they are synthesized.

Vesicles pinch off from the Golgi complex and carry the enclosed particles to the interior surface of the cell membrane, where the vesicle membrane and plasma membrane fuse and the contents of the vesicle are discharged.

External   Located on or near the surface.

Extracellular fluid (ECF)   Fluid outside the cells; includes interstitial fluid (fluid between the cells) and blood plasma.

Facial   Pertaining to the face.

Facilitated diffusion   A substance not soluble in lipids is transported across a plasma membrane by combining with a transporter (carrier) protein. Does not require ATP.

Fat-soluble vitamins   Vitamins A, D, E, and K. Classified as lipids because they are soluble in nonpolar solvents and insoluble in water.

Fatty acid   A hydrocarbon chain with a carboxyl group at one end. The hydrocarbon chain is nonpolar; the carboxyl group (—COOH) can function as an acid by releasing a hydrogen ion. Three fatty acids linked to a molecule of glycerol form a triglyceride (neutral fat).

Feedback system   A sequence of events in which information about the status of a specific aspect of the environment is continually reported (fed back) to a central control region. The three main components of a feedback system are : receptor (monitors changes), control center (decides how to respond), and effector (responds).

Femoral   Pertains to the thigh; region between the groin and the knee.

Fibroblast   A large, flat cell that forms the collagen fibers, elastic fibers, and matrix (intercellular substance) of loose connective tissue.

Fibrocyte   A mature fibroblast that no longer produces fibers or intercellular substances in connective tissue.

Filtration   The passage of a liquid through a filter or membrane. Blood plasma filters through pores (gaps between the cells lining walls of capillaries) into the spaces surrounding tissue cells.

Flagellum (plural: flagella)   A long, hairlike process on the outside of a spermatozoon; used for moving the cell.

Fluid compartments   The body fluids are separated into three main compartments: intracellular fluid (fluid inside the cells), interstitial fluid (fluid immediately surrounding the body cells), and plasma (fluid portion of the blood). The interstitial fluid and plasma together are called the extracellular fluid because they are outside the cells.

Fluid mosaic model   A model for plasma membrane structure; it suggests that plasma membranes consist of proteins floating like icebergs in a sea of lipids.

Foot   The terminal part of the lower extremity; the skeleton of the foot has 3 regions : the ankle (tarsus), arch or sole (metatarsus), and toes (phalanges).

Forearm   The part of the upper extremity between the elbow and the wrist.

Frontal plane   A plane that divides the body or an organ into anterior and posterior portions; it is at a right angle to the sagittal plane. Also called the *coronal plane*.

Gamete   A male or female reproductive cell; the spermatozoon (sperm) or ovum (egg).

Gene   A unit of hereditary information located in a definite position on a particular chromosome. The portion of a molecule of DNA that contains the nucleotide base sequence that codes for the sequence of amino acids in a particular polypeptide.

Gland Single or group of specialized epithelial cells that secrete substances such as hormones, mucus, oil, sweat, and digestive enzymes. The two main types of glands are exocrine and endocrine.

Glucose A six-carbon sugar, $C_6H_{12}O_6$; the major energy source for every cell type. Every known living cell breaks down (catabolizes) glucose for the production of ATP.

Glycogen A highly branched polymer of glucose containing thousands of subunits; stored in liver and muscle cells as an energy reserve.

Glycolipid A lipid molecule with a covalently bound carbohydrate group. The lipid portion is insoluble in water, while the carbohydrate portion is soluble.

Glycolysis The breakdown of a molecule of glucose into two molecules of pyruvic acid, yielding 2 ATP. Also called *anaerobic respiration*, because it requires no oxygen.

Glycoprotein A protein molecule with a covalently bound carbohydrate group. Most integral proteins in plasma membranes are glycoproteins.

Goblet cell A goblet-shaped unicellular gland that secretes mucus. Also called a *mucous cell*.

Golgi complex An organelle in the cytoplasm of cells consisting of four to eight flattened channels, stacked on one another, with expanded areas at their ends; its functions include packaging and secreting proteins, lipid secretion, and carbohydrate synthesis.

Gradient A gradient determines direction of movement, which is always from high concentration or pressure to low. Types of gradients include concentration gradients, pressure gradients, and electrical gradients.

Groin The depression between the thigh and the trunk; the inguinal region.

Ground substance The material that surrounds cells and fibers in connective tissue. It is secreted by fibroblasts and can be watery or gel-like. Ground substance together with the fibers constitute the matrix of connective tissues.

Hair A threadlike structure produced by a hair follicle that develops in the dermis. Also called *pilus*.

Hair follicle Structure composed of epithelium surrounding the root of a hair.

Hair root plexus A network of nerve endings arranged around the root of a hair that is stimulated when a hair shaft is moved. Functions as a touch receptor.

Hand The terminal portion of an upper extremity; includes the wrist (carpus), palm (metacarpus), and fingers (phalanges).

Haploid Having half the number of chromosomes characteristically found in the somatic (body) cells of an organism; the number of chromosomes in mature gametes (sperms and eggs). 23 chromosomes in humans. Symbolized by *n*.

Haversian canal *See* Central canal.

Haversian system *See* Osteon.

Hemoglobin A protein in red blood cells involved in the transport of oxygen and carbon dioxide.

Histology Microscopic study of the structure of tissues.

Homeostasis The relative stability of the internal environment (extracellular fluid). It results from the actions of feedback systems (reflex arcs), which constantly monitor changes and make adjustments by negative feedback.

Homologous chromosomes Two chromosomes that belong to a pair. They contain genes that control the same traits, but the genes are not always identical. One chromosome has been contributed by the father and the other by the mother. Also called *homologues*.

Hormone A secretion of an endocrine cell. It is released into the bloodstream, combines with the receptors of specific target cells, and alters a specific cell function.

Hydrogen bond A weak electrostatic attraction between partially charged atoms on adjacent molecules (or parts of the same large molecule). Most frequently, a positively polarized hydrogen atom is attracted to a negatively polarized oxygen or nitrogen atom.

Hydrolysis The breaking of a chemical bond with the addition of a water molecule.

Hypertonic A solution containing a higher concentration of solute than the intracellular fluid. There is a net movement of water molecules out of the cell by osmosis.

Hyponychium Free edge of the fingernail.

Hypotonic A solution containing a lower concentration of solute than the intracellular fluid. There is a net movement of water molecules into the cell by osmosis.

Inclusion Principally organic substance produced by a cell that is not enclosed by a membrane and may appear or disappear at various times in the life of a cell. Examples include glycogen, triglycerides, and melanin.

Inferior Away from the head or toward the lower part of a structure. Also called *caudad*.

Inguinal Pertaining to the groin (the depression between the thigh and the trunk).

Inorganic compound One of the two main types of chemical compounds. Inorganic compounds usually lack carbon; usually are small and contain ionic bonds. Examples include water and many acids, bases, and salts.

Integral protein Protein embedded in the phospholipid bilayer of plasma membranes; may span the entire membrane or be located on just one side.

Integumentary Pertaining to the skin.

Intercalated disc An irregular transverse thickening of plasma membrane between attached heart muscle cells. Contains desmosomes and gap junctions. Desmosomes hold cardiac muscle cells together; gap junctions aid in the conduction of muscle action potentials.

Intermediate Between two structures.

Intermediate filaments Protein filaments present in cells, ranging in diameter from 8 to 12 nanometers (nm). They give shape to a cell and provide structural reinforcement.

Internal Within the body.

Internal environment The extracellular fluid. The fluid outside the body cells; its two subdivisions are the interstitial fluid and the blood plasma.

Interphase The period of time between two mitotic divisions. During this phase each chromosome makes a copy of itself (replication) and materials needed by the cell are synthesized. Also called the *metabolic phase*.

Interstitial fluid The portion of extracellular fluid that fills the microscopic spaces between the cells of tissues. Also called *intercellular fluid* and *tissue fluid*.

Intracellular fluid (ICF) Fluid located within cells. Includes the cytosol, the fluid in the nucleus, and the fluid in other organelles.

Ion Any atom or small molecule that has a net positive or negative charge. Examples : Na$^+$ (sodium ion), Cl$^-$ (chloride ion), and HCO$_3^-$ (bicarbonate ion).

Ionic bond An electrostatic attraction between two oppositely charged ions.

Isotonic Having equal tension or tone. Having equal osmotic pressure between two different solutions. An isotonic solution has the same water concentration as the cytosol.

Isotopes Two or more atoms of the same element, each of which has the same number of protons in its nucleus, but a different number of neutrons. For example, an atom of carbon-12 has 6 protons and 6 neutrons; an atom of carbon-14 has 6 protons and 8 neutrons.

Keratin An insoluble protein found in hair, nails, and other keratinized tissues of the epidermis.

Keratinocyte The most numerous of the epidermal cells. Produces keratin.

Ketone bodies Products of fatty acid catabolism that accumulate in the blood during starvation and in untreated severe diabetes mellitus. They lower the pH.

Kinetic energy The energy of motion.

Krebs cycle A series of chemical reactions that occur in the matrix of mitochondria. Organic fragments derived from the breakdown of carbohydrates, proteins, and fats are catabolized to yield carbon dioxide, water, hydrogen atoms, and a small amount of ATP. The hydrogen atoms are transported by carrier molecules to the cristae of the mitochondria, where another series of reactions, called the electron transport system, yield carbon dioxide, water and a large number of ATP molecules. Also called the *citric acid cycle* and the *tricarboxylic acid cycle*.

Lacuna A small, hollow space found in compact bone tissue; location of mature bone cells called osteocytes.

Lamellae Concentric rings of hard, calcified matrix found in compact bone.

Lamellated corpuscle Oval-shaped pressure receptor located in subcutaneous tissue and consisting of concentric layers of connective tissue wrapped around a sensory nerve fiber. Also called a *Pacinian corpuscle*.

Lamina propria The connective tissue layer of a mucous membrane.

Langerhans cell A type of cell in the skin epidermis. It interacts with white blood cells in immune responses.

Lateral Farther from the midline of the body or a structure.

Leukocyte A white blood cell.

Ligand Any molecule or ion that binds to the surface of a protein by noncovalent bonds. All chemical messengers, such as hormones and neurotransmitters, are ligands.

Lipid (*lipos* = fat) An organic compound composed of carbon, hydrogen, and oxygen. It is usually insoluble in water, but soluble in alcohol, ether, and chloroform. Basic types of lipids are triglycerides, phospholipids, steroids, eicosanoids, fat-soluble vitamins, and carotenes.

Lipoprotein Aggregates of lipid molecules which are partially coated by proteins. Involved in the transport of lipids in the blood. High levels of low-density lipoproteins (LDL) are associated with increased risk of atherosclerosis; high levels of high-density lipoproteins (HDL) are associ-

ated with decreased risk of atherosclerosis.

Loose connective tissue A class of connective tissues in which the fibers are loosely woven and there are many cells. Types of loose connective tissues include: areolar, adipose, and reticular.

Lower extremity The appendage attached at the pelvic (hip) girdle, consisting of the thigh, knee, leg, ankle, foot, and toes.

Lumbar Region of the back and side between the ribs and pelvis; the loin.

Lunula The moon-shaped white area at the base of a nail.

Lysosome An organelle in the cytoplasm of a cell, enclosed by a single membrane and containing powerful digestive enzymes.

Macrophage A cell type which functions as a phagocyte in many tissues, eating bacteria and foreign materials. They have several other functions related to the immune responses of the body.

Mammary Pertaining to the breast.

Manual Pertaining to the hand.

Mast cell A cell found in areolar connective tissue along blood vessels. Local injury triggers mast cells to release histamine, a dilator of small blood vessels.

Matrix The ground substance and fibers that surround the cells in connective tissues.

Medial Nearer the midline of the body or a structure.

Mediastinum The mass of tissues between the lungs; it extends from the sternum to the vertebral column.

Meiosis The part of reproductive cell division that is concerned with the division of the nucleus. It results in the production of gametes (sex cells). Two successive nuclear divisions produce four daughter cells (gametes or sex cells) with the haploid number of chromosomes. (Division of the cytoplasm is called cytokinesis.)

Meissner's corpuscle *See* Corpuscle of touch.

Melanin A dark black, brown, or yellow pigment found in some parts of the the body such as skin and hair.

Melanoblast Precursor cell in the epidermis that gives rise to melanocytes, which produce melanin.

Melanocyte A pigmented cell located between or beneath cells of the deepest layer of the epidermis; synthesizes melanin.

Membrane A thin, flexible sheet of tissue. An *epithelial membrane* is composed of an epithelial layer and an underlying connective tissue layer; a *synovial membrane* is composed of areolar connective tissue only.

Merkel cell A type of cell located in the deepest layer of the epidermis of hairless skin. Makes contact with flattened sensory neuron endings called tactile discs; functions as a touch receptor.

Mesenchyme An embryonic connective tissue from which all other connective tissues arise.

Mesentery A fold of peritoneum attaching the small intestine to the posterior abdominal wall.

Mesoderm The middle of the three primary germ layers; gives rise to connective tissues and muscles.

Messenger RNA (mRNA) A single strand of nucleotides that carries the genetic code for a particular polypeptide from the nucleus to a ribosome. The sequence of nucleotide bases on the messenger RNA determines the sequence of amino acids in the polypeptide being synthesized.

Metabolic pathway   A sequence of enzyme-mediated reactions in a cell. In catabolic pathways large molecules are broken into smaller parts; in anabolic pathways new molecules are synthesized.

Metabolism   The sum of all the biochemical reactions that occur within an organism, including the synthetic (anabolic) reactions and decomposition (catabolic) reactions.

Metabolite   Any substance involved in a metabolic pathway.

Metacarpal   Pertaining to the metacarpus (the five bones of the hand between the wrist and the fingers); the palm.

Metaphase   The second stage of mitosis. Chromatid pairs line up on the metaphase plate of the dividing cell.

Metatarsal   Pertaining to the metatarsus (the five bones on the anterior portion of the foot between the instep and the toes).

Microfilament   Rodlike, protein filament about 6 nanometers (nm) in diameter. Contractile unit in muscle cells; provides support, shape, and movement in nonmuscle cells.

Microglia   Cells located in the brain and spinal cord that carry on phagocytosis. Also called *brain macrophages*.

Microtubule   Cylindrical protein filament; ranges in diameter from 18 to 30 nanometers (nm) and consists of the protein tubulin. Provides support, structure, and transportation inside cells.

Midline   An imaginary vertical line that divides the body into equal left and right sides.

Midsagittal plane   A vertical plane through the midline of the body or an organ; it divides the body or an organ into equal right and left sides. Also called a *median plane*.

Mineral   Inorganic substance (contains no carbon). Major minerals found in the body include calcium, sodium, potassium, iron, phosphorus, magnesium, and chloride.

Mitochondrion   A double-membraned organelle where nearly all of the cell's ATP is produced; known as the "powerhouse" of the cell.

Mitosis   The part of somatic cell division that is concerned with the division of the nucleus. It results in the production of two daughter nuclei that have the same number and kind of chromosomes as the original parent nucleus. (Division of the cytoplasm is called cytokinesis.)

Mitotic spindle   Collective term for a football-shaped assembly of microtubules that is responsible for the movement of chromosomes during cell division.

Mole   The unit of measure that is used for very large numbers of microscopic particles; 1 mole = 6.02×10^{23} particles.

Molecule   Two or more atoms in definite proportions linked by chemical bonds. A molecule of water consists of one oxygen atom chemically linked to two hydrogen atoms.

Monosaccharide   Single sugar. Examples include glucose, fructose, and galactose.

Mucous cell   *See* Goblet cell.

Mucous membrane   A membrane lining a body cavity that opens directly to the exterior. Also called the *mucosa*.

Mucus   The thick fluid secretion of mucous glands and mucous membranes.

Muscle tissue   Tissues specialized for contraction. There are three types of muscle tissues: skeletal, smooth, and cardiac.

Myofibril   Longitudinal bundle of thick and thin filaments located in the cytoplasm of a skeletal muscle fiber.

Nail   A hard plate, composed largely of keratinized cells of the epidermis. It develops from a region of epithelium under the nail root. It forms a protective covering on the dorsal surface of the distal phalanges of the fingers and toes.

Nail matrix   The region of epithelium under the nail root from which the nail is produced.

Negative feedback   The principle governing most control systems; a mechanism of response in which a stimulus initiates actions that reverse or reduce the stimulus.

Nervous tissue   Nervous tissue consists of two cell-type classifications: neurons (nerve cells) initiate and transmit nerve impulses; neuroglia insulate, nourish, support, and protect the neurons.

Neuroglia   Cells of the nervous system that are specialized to perform the functions of connective tissue. Neuroglia of the central nervous system include astrocytes, oligodendrocytes, microglia, and ependymal cells; neuroglia of the peripheral nervous system include the neurolemmocytes (Schwann cells) and the satellite cells. Also called *glial cells*.

Neuron   A nerve cell, consisting of a cell body, dendrites, and an axon.

Neutral fat   *See* Triglyceride.

Nucleic acid   An organic compound that consists of a long chain of nucleotides; each nucleotide contains a 5-carbon sugar, a phosphate group, and one of four possible nitrogenous bases: DNA nucleotides contain adenine, cytosine, guanine, or thymine; RNA nucleotides contain adenine, cytosine, guanine, or uracil.

Nucleoli (singular : nucleolus)   One or more spherical bodies in the nucleus. Units of ribosomes are assembled on the nucleoli then transported out of the nucleus into the cytosol.

Nucleus (plural : nuclei)   A prominent, spherical or oval cell organelle. It contains genes and controls cellular functions. (In the nervous system the term "nucleus" signifies a cluster of neuron cell bodies outside the brain or spinal cord.)

Oblique plane   A plane that passes through the body or an organ between transverse and frontal planes or between transverse and sagittal planes.

Oligodendrocyte   A neuroglial cell that produces a myelin sheath around axons in the central nervous system.

Organ   A structure with a specific function and usually a recognizable shape; composed of two or more different kinds of tissues.

Organelle   A permanent structure within a cell with a characteristic shape; it is specialized to serve a specific function in cellular activities.

Organic compound   A compound that contains carbon and hydrogen. Many organic compounds contain oxygen and nitrogen; some contain sulfur and phosphorus. Examples include carbohydrates, lipids, proteins, and nucleic acids (DNA and RNA).

Organism   A total living form; one individual.

Osmosis   The net movement of water molecules through a selectively permeable membrane from an area of high water concentration to an area of lower water concentration until an equilibrium is reached.

Osteon   The basic unit of structure in compact bones. Consists of a central canal with concentric rings of matrix, small spaces containing osteocytes, and small canals that carry nutrients and wastes. Also called an *Haversian system*.

Oxidation   The removal of hydrogen atoms (hydrogen ions and electrons) from a molecule. Or, less commonly, the addition of oxygen to a molecule that results in a decrease in the energy content of the molecule. The oxidation of glucose in the body is also called *cellular respiration*.

Pacinian corpuscle   *See* Lamellated corpuscle.

Papilla   A small nipple-shaped projection or elevation.

Papillary region   Outer 1/5 of the dermis; consists of areolar connective tissue with fine elastic fibers.

Parasagittal plane   A vertical plane that does not pass through the midline and that divides the body or an organ into unequal left and right portions.

Parietal   Pertaining to or forming the outer wall of a body cavity.

Parietal peritoneum   The portion of the peritoneum that lines the wall of the abdominopelvic cavity.

Parietal pleura   The outer layer of the membrane that encloses and protects the lungs; it is attached to the wall of the pleural cavity.

Patellar   Pertaining to the patella (kneecap).

Pectoral   Pertaining to the chest or breast.

Pedal   Pertaining to the foot.

Pelvic cavity   Inferior portion of the abdominopelvic cavity. It contains the urinary bladder, sigmoid colon, rectum, and internal female or male reproductive structures.

Pelvis   In the skeletal system, the pelvis is the basinlike structure formed by the two hipbones, the sacrum, and the coccyx. In the urinary system, the pelvis is the expanded, proximal portion of the ureter, lying within the kidney.

Pericardial cavity   Small, fluid-filled space surrounding the heart. Located between the parietal and visceral layers of the serous pericardium.

Pericardium   A loose-fitting membrane that encloses the heart. It consists of an outer fibrous layer and an inner serous layer.

Peripheral protein   A protein loosely attached to the inner or outer surface of a plasma membrane.

Peritoneum   The serous membrane that lines the abdominal cavity and covers the viscera (internal organs).

Peroxisome   Cell organelle similar in structure to a lysosome, but smaller. It contains enzymes for the production of hydrogen peroxide. Another enzyme present in the peroxisome uses the hydrogen peroxide to detoxify potentially harmful substances, such as alcohol, formaldehyde, and formic acid. Peroxisomes are abundant in liver cells.

pH   A symbol for the concentration of hydrogen ions in a solution. The pH scale extends from 0 to 14. A pH of 7 indicates neutrality; values less than 7 indicate increasing acidity; values higher than 7 indicate increasing alkalinity.

Phagocytosis   The process by which cells (phagocytes) ingest particulate matter; especially the ingestion and destruction of microbes, cell debris, and other foreign matter.

Phospholipid   A class of lipid that is a major component of plasma membranes. Consists of two fatty acids and a phosphate group attached to glycerol. The fatty acid "tails" are insoluble in water (hydrophobic) and the phosphate "heads" are soluble in water (hydrophilic).

Phospholipid bilayer   Arrangement of phospholipid molecules in two parallel rows in which the hydrophilic "heads" face outward and the hydrophobic "tails" face inward. The lipid portion of a plasma membrane has this structure.

Phosphorylation   The addition of a phosphate group to an organic molecule.

Physiology   Science that deals with the functions of an organism.

Pilus   (plural : pili)   A hair.

Pinocytosis   The process by which cells ingest liquid.

Plasma   The extracellular fluid found in blood vessels; blood minus the blood cells and platelets.

Plasma membrane   The outer, limiting membrane that separates the cell's internal parts from the interstitial fluid. Also called the *cell membrane*.

Pleura   The serous membrane that covers the lungs and lines the walls of the chest and the diaphragm.

Pleural cavity   A small, fluid-filled space that surrounds each lung; the space between the visceral and parietal pleurae.

Polar covalent bond   A covalent bond in which two electrons are shared unequally between two atoms. The more electronegative atom pulls the electrons closer. An example is an O—H bond; the oxygen atom is more electronegative, so the covalent bond is negative near the oxygen and positive near the hydrogen.

Polarized   A condition in which opposite effects or states exist at the same time. For example, a cell membrane is electrically polarized; its outer surface is positively charged and its inner surface is negatively charged.

Pollex   The thumb.

Polymer   A large molecule formed by linking together many smaller subunits (called monomers). Examples include: polysaccharides (monomers are glucose molecules); proteins (monomers are amino acids); nucleic acids (monomers are nucleotides).

Polypeptide   A chain of amino acids; anywhere from ten amino acids to over two thousand.

Polysaccharide   A carbohydrate in which three or more monosaccharides are joined chemically.

Polyunsaturated fat   A fatty acid that contains more than one double covalent bond between its carbon atoms; abundant in triglycerides of corn oil, safflower oil, and cottenseed oil.

Posterior   Nearer to the back; opposite of anterior. Also called *dorsal*.

Potential energy   Inactive or stored energy.

Primary germ layers   The three layers of embryonic tissue that give rise to all tissues of the mature organism. They include the ectoderm, mesoderm, and endoderm.

Product   The molecules formed in an enzyme-catalyzed chemical reaction.

Prophase   The first stage of mitosis during which chromatin shortens and coils into visible chromosomes. The chromosomes have been replicated during interphase; two identical chromosomes joined by a centromere are called chromatids.

Prostaglandin (PG)   A class of lipids derived from arachidonic acid (a 20-carbon fatty acid). Prostaglandins function mainly as local hormones (hormones that act on the same cells that secrete them or on nearby cells).

Protein   An organic compound consisting of carbon, hydrogen, oxygen, nitrogen, and sometimes sulfur and phosphorus. Proteins consist of one to several polypeptide chains.

Protein kinase   An enzyme which catalyzes the addition of a phosphate group to a specific protein.

Proximal   Nearer to the point of origin; away from the hand of the upper extremity or away from the foot of the lower extremity.

Pyruvic acid   A 3-carbon molecule that is the end-product of glycolysis. When oxygen is present, pyruvic acid enters a mitochondrion and is converted into acetyl coenzyme A, which enters the Krebs cycle. In the absence of oxygen, pyruvic acid is converted into lactic acid.

Quadrant   One of four parts.

Reactant   A molecule that chemically reacts with another molecule. In enzyme-catalyzed reactions, a reactant is called a substrate; it attaches to the active site of an enzyme.

Receptor   There are two basic types of receptors : In the sensory system, a receptor is a specialized peripheral ending of a sensory neuron (or a specialized cell closely associated with it) that detects changes in the environment. In terms of chemical communication, a receptor is a protein molecule with a binding site for a specific chemical messenger, such as a hormone.

Receptor-mediated endocytosis   A highly selective process in which cells take up large molecules or particles. In this process, substances bind to receptors on the plasma membrane, triggering an infolding of the membrane and the formation of an endocytic vesicle. The ingested particles are eventually broken down by enzymes in lysosomes.

Receptor modulation   When a ligand attaches to a protein receptor, the shape of the receptor is changed (modulated). If the protein receptor is also an enzyme, the change in shape may activate it or deactivate it.

Reduction   The addition of electrons and hydrogen ions (hydrogen atoms) to a molecule. Or, less commonly, the removal of oxygen from a molecule that results in an increase in the energy content of the molecule.

Reflex   Fast response to a change (stimulus) in the internal or external environment; a biological control system that links a stimulus with a response.

Reflex arc   The pathway followed by a reflex, composed of neural, hormonal, or a combination of neural and hormonal elements. There are three fundamental components of a reflex arc : a sensory receptor that detects a change in the environment, an integrating center that decides how to respond, and an effector (a muscle that contracts or a gland that secretes a substance). Also called *feedback response*.

Regional anatomy   The division of anatomy dealing with a specific region of the body, such as the head, neck, chest, or abdomen.

Reproductive cell division   Type of cell division in which sperm and egg cells are produced; consists of meiosis (nuclear division) and cytokinesis (cytoplasmic division).

Response   The change brought about by a feedback system (reflex arc). Responses involve the secretion of glands or the contraction of muscles.

Reticular region   The inner 4/5 of the skin dermis. Consists of dense, irregular connective tissue containing interlacing bundles of collagen and coarse elastic fibers.

Reticulum   A network.

Reversible reaction   A chemical reaction that includes both synthesis and decomposition reactions. The reaction can go in either direction, depending upon the conditions.

Ribonucleic acid (RNA)   A single-stranded nucleic acid constructed of nucleotides consisting of one of four possible nitrogenous bases (adenine, cytosine, guanine, or uracil), a 5-carbon sugar, and a phosphate group. There are three types: messenger RNA (mRNA), transfer RNA (tRNA), and ribosomal RNA (rRNA). All three types are involved in protein synthesis.

Ribosomes   Tiny, spherical cell organelles. Provide the sites for protein synthesis.

Sagittal plane   A vertical plane that divides the body or an organ into left and right portions. There are two types of sagittal planes : a midsagittal (median) plane divides the body into two equal halves; a parasagittal plane divides the body into unequal portions.

Saturated fat   A fatty acid that contains no double bonds between any of its carbon atoms; all are single bonds and all carbon atoms are saturated in the sense that they are bonded to the maximum number of hydrogen atoms. Found in foods of animal origin such as meat, milk products, milk, and eggs.

Sebaceous   Secreting oil.

Sebaceous gland   An exocrine gland usually associated with a hair follicle of the skin; secretes sebum. Also called an *oil gland*.

Secretion   Production and release of organic molecules, ions, and water by gland cells in response to a specific stimulus.

Selectively permeable membrane   A membrane that permits the passage of certain substances, but restricts the passage of others. Also called a *semipermeable membrane*.

Serosa   *See* Serous membrane.

Serous membrane   A membrane that lines a body cavity. Also called the *serosa*.

Sex chromosomes   The twenty-third pair of chromosomes, designated X and Y. They determine the sex of an individual: XX is female; XY is male.

Skeletal muscle   An organ specialized for contraction, composed of striated muscle fibers (cells), supported by connective tissue. A skeletal muscle is attached to a bone by a tendon (fibrous band) or an aponeurosis (fibrous sheet composed of collagenous bundles).

Skeletal muscle tissue   Cylindrical, striated fibers with many peripheral nuclei; under voluntary control.

Smooth muscle tissue   Spindle-shaped, nonstriated fibers with one centrally located nucleus; usually under involuntary control.

Sodium-potassium ATP-ase   An active transport system located in the cell membrane that transports sodium ions out of the cell and potassium ions into the cell at the expense of ATP. It keeps the ionic concentrations of these ions at the appropriate functional levels. Also called the *sodium pump*.

Solute   The molecules or ions dissolved in a liquid.

Solution   A liquid containing dissolved molecules or ions.

Solvent   The liquid in which molecules or ions are dissolved.

Somatic   Pertaining to the body, especially the outer walls and framework of the body (skin, skeletal muscles, tendons, and joints). Somatic cells include all cells of the body except gametes (sperm and eggs).

Somatic cell division   Type of cell division in which a single starting cell (parent cell) duplicates itself to produce two identical cells (daughter cells). Consists of mitosis (nuclear division) and cytokinesis (cytoplasmic division).

Squamous   Scalelike.

Starch   A branched polysaccharide chain composed of thousands of glucose subunits. Starch is found in plant cells; the corresponding polysaccharide found in animal cells, especially liver and muscle cells, is called glycogen.

Plants store chemical energy in the form of starch.

Steroid A class of lipid. Consists of a skeleton of four carbon rings to which various polar groups may be attached. Sex hormones and hormones produced by the adrenal glands are steroids.

Stimulus Any detectable change in the internal or external environment.

Stratum A layer.

Subcutaneous Beneath the skin. Also called *hypodermic*.

Subcutaneous layer A continuous sheet of areolar connective tissue and adipose tissue between the dermis of the skin and the deep fascia of the muscles. Also called the *superficial fascia*.

Subserous fascia A layer of connective tissue lying between the deep fascia and the serous membrane lining a body cavity.

Substrate A substance to which an enzyme attaches during an enzyme-catalyzed chemical reaction.

Sudoriferous gland A gland in the dermis or subcutaneous layer that produces perspiration. Also called a *sweat gland*.

Superficial Located on or near the surface of the body.

Superficial fascia A continuous sheet of fibrous connective tissue between the dermis of the skin and the deep fascia of the muscles. Also called *subcutaneous layer*.

Superior Toward the head or upper part of the body or a structure. Also called *cephalad* or *craniad*.

Symport Process by which two substances, usually Na$^+$ and another substance, move in the same direction in a plasma membrane. Also called *cotransport*.

Synapsis The coming together of homologous chromatids during prophase of the first meiotic division. Homologous chromatids have genes that control the same traits, but they are not identical. Two homologous chromatids may break and exchange portions (crossing-over) during synapsis; the result is a new combination of genetic information for the chromatids involved.

Synthesis reaction A reaction in which reactants combine to form a new product; chemical bonds are formed.

System An association of organs that have a common function.

Systemic anatomy The study of particular systems of the body, such as the skeletal, muscular, nervous, cardiovascular, or urinary systems.

Tarsal Pertaining to the tarsus.

Tarsus A collective term for the seven bones of the ankle.

Telophase The final stage of mitosis. Nuclear envelope reappears and encloses chromosomes; chromosomes resume chromatin form; nucleoli reappear; mitotic spindle disappears.

Thermoreceptor Sensory receptor in the skin that detects changes in temperature.

Thoracic Pertaining to the thorax or chest.

Thoracic cavity Superior component of the ventral cavity. It contains the two pleural cavities surrounding the lungs and the collection of tissues called the mediastinum between the lungs.

Tissue A group of similar cells and their intercellular substance joined together to perform a specific function.

Transcription Messenger RNA synthesis. During protein synthesis, DNA base sequences are "rewritten" as messenger RNA base sequences. Allows genetic information to be carried from the nucleus to a ribosome.

Transfer RNA A strand of RNA shaped like a clover leaf that carries a specific amino acid to a particular codon on a messenger RNA.

Translation Polypeptide synthesis. During protein synthesis, messenger RNA base sequences are "translated" into the corresponding amino acid sequences. The result is the formation of a polypeptide with a specific amino acid sequence.

Transverse plane A plane that divides the body or an organ into superior and inferior portions. Also called *cross-sectional planes* and *horizontal planes*.

Triglyceride A lipid formed from one molecule of glycerol and three molecules of fatty acids. Triglycerides are the body's most highly concentrated form of stored chemical energy; a given weight of triglyceride contains over 2 times as much chemical energy as an equal weight of carbohydrate or protein. Triglycerides are stored in fat cells (adipocytes). Also called a *neutral fat*.

Trunk The part of the body to which the upper and lower extremities are attached. Also called the *torso*.

Umbilical Pertaining to the umbilicus or navel.

Umbilicus A small scar on the abdomen that marks the former attachment of the umbilical cord to the fetus. Also called the *navel*.

Upper extremity The appendage attached at the shoulder girdle, consisting of the arm, forearm, wrist, hand, and fingers.

Valence electrons The electrons in the outer shell of an atom that must be lost, gained, or shared when a chemical bond is formed with another atom.

Ventral Nearer to the front; opposite of dorsal. Also called *anterior*.

Ventral body cavity Cavity near the front of the body that contains viscera (internal organs) and consists of two subdivisions: a superior thoracic cavity and an inferior abdominopelvic cavity.

Vertebral canal One of the two main cavities that make up the dorsal body cavity. It is located within the vertebral column and contains the spinal cord. Also called the *spinal canal*.

Vesicle A small bladder or sac containing liquid.

Viscera (singular: viscus) The organs inside the ventral body cavity.

Visceral Pertaining to the organs or to the coverings of the organs of the ventral body cavity.

Visceral peritoneum The portion of the peritoneum that covers the viscera (internal organs of the abdominopelvic cavity).

Visceral pleura The inner layer of the serous membrane that covers the lungs.

Bibliography

Curtis, Helena.   *Biology,* 3rd ed.
New York : Worth, 1979.

Dorland, William Alexander.   *Dorland's Illustrated Medical Dictionary,* 27th ed.
Philadelphia : W. B. Saunders, 1988.

Ganong, William F.   *Review of Medical Physiology,* 15th ed.
Norwalk, Connecticut : Appleton & Lange, 1991.

Junqueira, L. Carlos, Jose Carneiro, and Robert O. Kelley.   *Basic Histology,* 6th ed.
Norwalk, Connecticut : Appleton & Lange, 1989.

Kapit, Wynn and Lawrence M. Elson.   *The Anatomy Coloring Book.*
New York : Harper & Row, 1977.

Kimball, John W.   *Biology,* 4th ed.
Reading, Massachusetts : Addison-Wesley, 1978.

Melloni, B. J., Ida Dox, and Gilbert Eisner.   *Melloni's Illustrated Medical Dictionary,* 2nd ed.
Baltimore : Williams & Wilkins, 1985.

Moore, Keith L.   *Clinically Oriented Anatomy,* 3rd ed.
Baltimore : Williams & Wilkins, 1992.

Roberts, M. B. V.   *Biology : A Functional Approach,* 2nd ed.
Sudbury-on-Thames, England : Thomas Nelson and Sons Ltd., 1976.

Tortora, Gerard J. and Sandra Reynolds Grabowski.   *Principles of Anatomy and Physiology,* 7th ed.
New York : HarperCollins, 1993.

Vander, Arthur J., James H. Sherman, and Dorothy S. Luciano.   *Human Physiology,* 5th ed.
New York : McGraw-Hill, 1990.